Advances in Science, Technology & Innovation

IEREK Interdisciplinary Series for Sustainable Development

Advances in Science, Technology & Innovation (ASTI) is a series of peer-reviewed books based on the best studies on emerging research that redefines existing disciplinary boundaries in science, technology and innovation (STI) in order to develop integrated concepts for sustainable development. The series is mainly based on the best research papers from various IEREK and other international conferences, and is intended to promote the creation and development of viable solutions for a sustainable future and a positive societal transformation with the help of integrated and innovative science-based approaches. Offering interdisciplinary coverage, the series presents innovative approaches and highlights how they can best support both the economic and sustainable development for the welfare of all societies. In particular, the series includes conceptual and empirical contributions from different interrelated fields of science, technology and innovation that focus on providing practical solutions to ensure food, water and energy security. It also presents new case studies offering concrete examples of how to resolve sustainable urbanization and environmental issues. The series is addressed to professionals in research and teaching, consultancies and industry, and government and international organizations. Published in collaboration with IEREK, the ASTI series will acquaint readers with essential new studies in STI for sustainable development.

More information about this series at http://www.springer.com/series/15883

Haroun Chenchouni • Ezzoura Errami
Fernando Rocha • Luisa Sabato
Editors

Exploring the Nexus of Geoecology, Geography, Geoarcheology and Geotourism: Advances and Applications for Sustainable Development in Environmental Sciences and Agroforestry Research

Proceedings of the 1st Springer Conference of the Arabian Journal of Geosciences (CAJG-1), Tunisia 2018

Springer

Editors
Haroun Chenchouni
University of Tebessa
Tebessa, Algeria

Ezzoura Errami
Chouaïb Doukkali University
El Jadida, Morocco

Fernando Rocha
Department of Geosciences
University of Aveiro
Aveiro, Portugal

Luisa Sabato
Università degli Studi di Bari
Aldo Moro
Bari, Italy

ISSN 2522-8714 ISSN 2522-8722 (electronic)
Advances in Science, Technology & Innovation
IEREK Interdisciplinary Series for Sustainable Development
ISBN 978-3-030-01682-1 ISBN 978-3-030-01683-8 (eBook)
https://doi.org/10.1007/978-3-030-01683-8

Library of Congress Control Number: 2018958931

This Springer imprint is published by the registered company Springer Nature Switzerland AG
The registered company address is: Gewerbestrasse 11, 6330 Cham, Switzerland

Preface

The valuation of ecosystem functions, goods and services of natural and cultural heritage and bioresources of many regions worldwide is still little known. The Mediterranean, Middle East region as well as the surrounding area are ecologically unique due to large differences in climatic, geographical and geological features, which offer countries of these region weighty socio-economic potentials in terms of fertile agricultural lands, rich natural resources and the existence of strategic resources such as crude oil and natural gas. Environmental degradation is one of the largest threats that are being looked at in the world today, and this specific region is no exception. Inappropriate and adverse human practices and activities such as disturbances land damage, different forms of pollutions, overpopulation, landfills, deforestation, combined with natural drivers are the main causes of land degradation and ecosystem destruction. The destruction of the environment ravages large parts of the planet, in particular, heavily populated and/or exploited regions, and thus threatens the existence of all species, including ours. Furthermore, economic companies maximize their profits by ignoring environmental protection, while governments encourage investment rather than strict regulations. These challenging issues hamper the social welfare and sustainable development, which depends on three pillars of sustainability: 'Environment', 'Society' and 'Economy'. Public environmental consciousness, however, keeps increasing with the increase of major global problems such as climate change. Nowadays, people are interested in learning more about new advances in environmental research initiatives in view of the ever growing environmental degradation. If efforts of scholar ecologists and environmental activists are considered to have paid off socially, the state of the environment of our planet has not improved as looked-for, yet.

This proceedings volume is based on the accepted papers for both oral and poster presentations during the 1st Springer Conference of the Arabian Journal of Geosciences (CAJG-1), Tunisia 2018. The book offers a broad range of new studies that discuss latest advances in geoenvironmental sciences from diverse backgrounds including agroecology, climate change, environmental biotechnology, biodiversity, geotourism and geoarchaeology. It shares insights of experienced scholars from, but not limited to, research institutes in the Mediterranean and Middle East region on how the understanding of ecological processes is the key for improving practices in environment management and conservation. The main topics include Environmental Assessment and Monitoring, Agroforestry Systems-Environment interactions, Environmental Impacts, Restoration Ecology, Investigations and Applications of Environmental Biotechnology, Spatiotemporal Patterns of Biodiversity, Paleobiology, and Socio-economics of Geotourism and Archaeology. With its diverse topics and new results, this volume enhances the understanding of environmental impacts and the state of restoration ecology in natural and agricultural habitats as it maximizes the readers' insights into emerging environmental issues and challenges. It provides new insights and practical information based on the latest data reconstructed from (i) various environmental assessments and monitoring of agrisilvicultural systems, (ii) applications of environmental biotechnology in plant breeding and plant-product valuation, (iii) spatiotemporal patterns of biodiversity and terrestrial paleobiodiversity, and (iv) socio-economic assessments of sustainable development involving geotourism and archaeological sites.

Tebessa, Algeria	Haroun Chenchouni
El Jadida, Morocco	Ezzoura Errami
Aveiro, Portugal	Fernando Rocha
Bari, Italy	Luisa Sabato
July 2018	

Acknowledgements

Our appreciation is extended to the authors of the papers for their hard and diligent work and producing high-quality contributions. We would like to thank the reviewers of the papers for their in-depth reviews and great efforts in improving the quality of the papers. Also, thanks are extended to Amjad Kallel who supervised and handled the evaluation process, to Sahbi Moalla who handled the submission and evaluation system for the ten conference proceedings volumes, and the publishing staff of Springer headed by Nabil Khélifi, Senior Editor for their efforts and contributions in completing this conference proceedings volume. All the above-mentioned efforts were very important in making this book a success.

About the 1st Springer Conference of the Arabian Journal of Geosciences (CAJG-1), Tunisia 2018

The *Arabian Journal of Geosciences (AJG)* is a Springer journal publishing original articles on the entire range of Earth sciences in partnership with the Saudi Society for Geosciences. The journal focuses on, but not limited to, research themes which have regional significance to the Middle East, the Euro-Mediterranean, Africa, and Asia. The journal receives on average 2000 submissions a year and accepts around 500 papers for publication in its 24 annual issues (acceptance rate 25%). It enjoys the participation of an editorial team of 100 international associate editors who generously help in evaluating and selecting the best papers.

In 2008, Prof. Abdullah Al-Amri, in close partnership with Springer, founded the Arabian Journal of Geosciences (AJGS). In this year, the journal celebrates its tenth anniversary. On this occasion and to mark this event, the Founder and Editor-in-Chief of the AJGS Prof. Al-Amri organized in close collaboration with Springer the 1st Conference of the Arabian Journal of Geosciences (1st CAJG) in Hammamet, Tunisia, from November 12 to 15, 2018 (www.cajg.org).

The conference was an occasion to endorse the journal's long-held reputation for bringing together leading authors from the Middle East, the Euro-Mediterranean, Africa, and Asia who work at the wide-ranging fields of Earth sciences. The conference covered all crosscutting themes of Geosciences and focused principally on the following ten tracks:

- Track 1. Climate, paleoclimate, and paleoenvironmental changes
- Track 2. Geoinformatics, remote sensing, geodesy
- Track 3. Geoenvironmental engineering, geomechanics and geotechnics, geohazards
- Track 4. Geography, geoecology, geoarcheology, geotourism
- Track 5. Geophysics, seismology
- Track 6. Hydrology, hydrogeology, hydrochemistry
- Track 7. Mineralogy, geochemistry, petrology, and volcanology
- Track 8. Petroleum engineering and petroleum geochemistry
- Track 9. Sedimentology, stratigraphy, paleontology, geomorphology, pedology
- Track 10. Structural/petroleum/mining geology, geodynamics, marine geology

The dynamic four-day conference provided more than 450 attendees with opportunities to share their latest unpublished findings and learn the newest geoscience studies. The event also allowed attendees to meet and discuss with the journal's editors and reviewers.

More than 950 short contributing papers to the conference were submitted by authors from more than 70 countries. After a pre-conference peer review process by more than 500 reviewers, 700 papers were accepted. These papers were published as chapters in the conference proceedings by Springer.

The conference proceedings consist of ten edited volumes, each edited by the following group of Arabian Journal of Geosciences (AJGS) editors and other guest editors:

Volume 1. Patterns and Mechanisms of Climate, Paleoclimate, and Paleoenvironmental Changes from Low-Latitude Regions

Zhihua Zhang (AJGS Editor): Beijing Normal University, Beijing, China
Nabil Khélifi (AJGS Editor): Earth Sciences Editorial Department, Springer, Heidelberg, Germany
Abdelkader Mezghani (Guest Editor): Norwegian Meteorological Institute, Norway
Essam Heggy (Guest Editor): University of Southern California and NASA Jet Propulsion Laboratory, Caltech, USA

Volume 2. Advances in Remote Sensing and Geo Informatics Applications

Hesham M. El-Askary (Guest Editor): Schmid College of Science and Technology at Chapman University, USA
Saro Lee (AJGS Editor): Korea Institute of Geoscience and Mineral Resources, Daejeon, South Korea
Essam Heggy (Guest Editor): University of Southern California and NASA Jet Propulsion Laboratory, Caltech, USA
Biswajeet Pradhan (AJGS Editor): University of Technology Sydney, Sydney, Australia

Volume 3. Recent Advances in Geo-Environmental Engineering, Geomechanics and Geotechnics, and Geohazards

Amjad Kallel (AJGS Editor): ENIS, University of Sfax, Tunisia
Zeynal Abiddin Erguler (AJGS Editor): Dumlupinar University, Kutahya, Turkey
Zhen-Dong Cui (AJGS Editor): China University of Mining and Technology, Xuzhou, Jiangsu, China
Ali Karrech (AJGS Editor): The University of Western Australia, Australia
Murat Karakus (AJGS Editor): University of Adelaide, Australia

Pinnaduwa Kulatilake (AJGS Editor): Department of Materials Science and Engineering, The University of Arizona, USA
Sanjay Kumar Shukla (AJGS Editor): School of Engineering, Edith Cowan University, Perth, Australia

Volume 4. Exploring the Nexus of Geoecology, Geography, Geoarcheology, and Geotourism: Advances and Applications for Sustainable Development in Environmental Sciences and Agroforestry Research

Haroun Chenchouni (AJGS Editor): University of Tebessa, Algeria
Ezzoura Errami (Guest Editor): Chouaïb Doukkali University, El Jadida, Morocco
Fernando Rocha (Guest Editor): University of Aveiro, Portugal
Luisa Sabato (AJGS Editor): Università degli Studi di Bari "Aldo Moro", Bari, Italy

Volume 5. On Significant Applications of Geophysical Methods

Narasimman Sundararajan (AJGS Editor): Sultan Qaboos University, Muscat, Oman
Mehdi Eshagh (AJGS Editor): University West, Trollhättan, Sweden
Hakim Saibi (AJGS Editor): United Arab Emirates University, Al-Ain, Abu Dhabi, UAE
Mustapha Meghraoui (AJGS Editor): Université de Strasbourg, Strasbourg, France
Mansour Al-Garni (AJGS Editor): King Abdulaziz University, Jeddah, Saudi Arabia
Bernard Giroux (AJGS Editor): Centre Eau Terre Environnement, Québec, Canada

Volume 6. Advances in Sustainable and Environmental Hydrology, Hydrogeology, Hydrochemistry and Water Resources

Helder I. Chaminé (AJGS Editor): School of Engineering—ISEP, Polytechnic of Porto, Portugal
Maurizio Barbieri (AJGS Editor): University of Rome La Sapienza, Italy
Ozgur Kisi (AJGS Editor): Ilila State University, Tbilisi, Georgia
Mingjie Chen (AJGS Editor): Sultan Qaboos University, Muscat, Oman
Broder J. Merkel (AJGS Editor): TU Bergakademie Freiberg, Freiberg, Germany

Volume 7. Petrogenesis and Exploration of the Earth's Interior

Domenico Doronzo (AJGS Editor): Consejo Superior de Investigaciones Cientificas, Spain
Emanuela Schingaro (AJGS Editor): Università degli Studi di Bari Aldo Moro—UniBa, Italy
John S. Armstrong-Altrin (AJGS Editor): The National Autonomous University of Mexico, Mexico
Basem Zoheir (Guest Editor): Benha University, Egypt and University of Kiel, Germany

Volume 8. Advances in Petroleum Engineering and Petroleum Geochemistry

Santanu Banerjee (AJGS Editor): Indian Institute of Technology Bombay, Mumbai, India
Reza Barati (AJGS Editor): The University of Kansas, Lawrence, KS, USA
Shirish Patil (Guest Editor): Saudi Aramco and King Fahd University of Petroleum and Minerals, Dhahran, Saudi Arabia

Volume 9. Paleobiodiversity and Tectono-sedimentary Records in the Mediterranean Tethys and Related Eastern Areas

Mabrouk Boughdiri (AJGS Editor): University of Carthage, Amilcar, Tunisia
Beatriz Bádenas (AJGS Editor): University of Zaragoza, Zaragoza, Spain
Paul Selden (AJGS Editor): University of Kansas, Lawrence, Kansas, USA
Etienne Jaillard (Guest Editor): Université Grenoble Alpes, France
Peter Bengtson (AJGS Editor): Universität Heidelberg, Heidelberg, Germany
Bruno R. C. Granier (AJGS Editor): Université de Bretagne Occidentale, Brest, France

Volume 10. The Structural Geology Contribution to the Africa-Eurasia Geology: Basement and Reservoir Structure, Ore Mineralisation and Tectonic Modelling

Federico Rossetti (Guest Editor): Università Roma Tre, Roma, Italy
Ana Crespo Blanc (Guest Editor): University of Granada, Spain
Federica Riguzzi (Guest Editor): National Institute of Geophysics and Volcanology, Roma, Italy
Estelle Leroux (Guest Editor): IFREMER, Unité Géosciences Marines, Plouzané, France
Kosmas Pavlopoulos (Guest Editor): Paris Sorbonne University Abu Dhabi, Abu Dhabi, UAE
Olivier Bellier (Guest Editor): CEREGE, Aix-en-Provence, France
Vasilios Kapsimalis (Guest Editor): Institute of Oceanography, Hellenic Centre for Marine Research, Anavyssos, Greece

About the Conference Steering Committee

General Chair

Abdullah Al-Amri: Founder and Editor-in-Chief of AJGS, King Saud University, Saudi Arabia

Conference Supervisor

Nabil Khélifi: Senior Publishing Editor, Springer Middle East and North African Program Springer, a part of Springer Nature, Heidelberg, Germany

Scientific Committee Chair

François Roure: Guest of Editorial Board of AJGS, IFP—Energies Nouvelles, France

Walter D. Mooney: Guest of Editorial Board of AJGS, US Geological Survey Western Region, USA

Local Organization Chair

Mabrouk Boughdiri: Associate Editor of AJGS, University of Carthage, Bizerte, Tunisia

Evaluation Chair

Amjad Kallel: Assistant Editor of AJGS, ENIS, University of Sfax, Tunisia

Publication Chair

Biswajeet Pradhan: Associate Editor of AJGS, University of Technology Sydney, Sydney, Australia

Essam Heggy: Guest of Editorial Board of AJGS, University of Southern California and NASA Jet Propulsion Laboratory, Caltech, USA

Program Chair

Hakim Saibi: Associate Editor/Assistant Editor of AJGS, United Arab Emirates University, Al-Ain, Abu Dhabi, UAE

Domenico Doronzo: Associate Editor/Assistant Editor of AJGS, Consejo Superior de Investigaciones Cientificas, Spain

Communication Chair

Mohamed Ksibi: Guest of Editorial Board of AJGS, ISBS, University of Sfax, Tunisia

English Language Advisory Committee

Abdelmajid Dammak: ENIS, University of Sfax, Tunisia
Chokri Khalaf: FMS, University of Sfax, Tunisia
Dhouha Mabrouk: FLSHS, University of Sfax, Tunisia
Mohamed Elbahi: ENIS, University of Sfax, Tunisia
Sami Shami: ENIS, University of Sfax, Tunisia
Yasmine Basha: FLSHS, University of Sfax, Tunisia

Conference Manager

Mohamed Sahbi Moalla: Coordinator of AJGS, ISET, University of Sfax, Tunisia

Contents

About the Editors

Dr. Haroun Chenchouni is Senior Lecturer and Research Ecologist in The Department of Natural and Life Sciences at The University of Tebessa (Algeria). He teaches Forest Ecology, Soil Ecology, Biostatistics and Ecological modelling. He graduated as an Engineer in Plant Ecology and Forest Ecosystems from the Department of Biological Sciences (University of Batna, Algeria). He holds an M.Sc. in Dryland Ecology from the University of Ouargla (Algeria) and a Doctoral Degree in Ecology from the University of Batna. His research interests are fairly broad, but revolve around biological interactions and community ecology of arid and semiarid lands of North African ecosystems. He is using statistical modelling approaches to understand how natural environments and anthropogenic perturbations affect biological interactions, shape trends in population dynamics and influence community diversity. He has published more than 50 papers in international journals, six proceedings in international conferences and two books. He is editor and/or editorial board member of many international journals. In 2017, he joined the AJGS as an Associate Editor responsible for evaluating submissions in the fields of Biogeosciences, Geoecology, Climate Change, Plant and Soil Science, Agricultural and Forest Environment, and Environmental Science.

Prof. Dr. Ezzoura Errami is a Moroccan geoscientist with more than 12 years of activities at the continental and international levels, during which she has been actively involved in bilateral cooperation with many international professional organizations of Earth sciences. She is currently working as a Full Professor at the Geology Department of Chouaïb Doukkali University in Morocco. There, she has also achieved in 2001 her Ph.D. project in collaboration with Ruprecht Karls University in Germany. Her research focuses on petrology, geochemistry, structural geology, geoheritage, geotourism, geoeducation, sustainable development and gender-related studies in geosciences. She is the author and co-author of many scientific papers, conferences papers, books, reports and guidebooks. She has chaired many organizing committees and/or scientific committees of international scientific events. She serves as AAWG assistant general secretary, president and AGN and DESAME coordinator, IUGS councillor, ArabU vice president, CIFEG Managing Board member and IAPG Africa coordinator.

Prof. Fernando Rocha is born in Lisbon on 7 January 1956. He achieved degree in Geology from the University of Lisbon, in 1984; Ph.D. in Geosciences (Geology) from the University of Aveiro, in 1994; and D.Sc. (Habilitation) in Geosciences (Clay Science) from the University of Aveiro, in 2000. He is Full Professor (Mineral Resources, Geosciences Department) since 1 August 2002 and Director of GEOBIOTEC Research Centre since 2007. He has played several positions at the University of Aveiro, including Head of the Department of Geosciences (1998–2002, 2011–2015), Pro-Rector (infrastructures and Campus management, 2003–2007) and Vice-Rector (research, innovation and technology transfer, 2008–2010). After a brief passage (1984/85) by the private sector of public works as Engineering Geologist, he developed academic career at the University of Aveiro since 1986, with scientific and pedagogical activity in the fields of mineral resources, marine and coastal geology, and medical and environmental geology.

Dr. Luisa Sabato is Associate Professor of Stratigraphy and Sedimentology at the University of Bari (Italy) and currently is the President of the Master degree in Geological Sciences.

She is tutor of several Ph.D. students and teaches Geology at the degree courses in Geological Sciences, Natural Sciences, Environmental Sciences, Sciences and management of maritime activities.

She is in the Management Board of the GeoSed (Italian Association for Sedimentary Geology) and SGI (Italian Geology Society).

She has chaired many scientific national and international geological meetings and organized national and international geological Congresses and field trips. She participated also as coordinator to projects regarding sedimentary geology and geosites, and is author and co-author of many scientific papers, geological maps and field trip guidebooks.

Her main research fields regard sedimentary geology, with particular attention to the facies analysis of continental and shallow marine systems. Other topics deal with problems of fluvial dynamics and human impact in alluvial and coastal plains. She also deals with geological mapping within the CARG Project (promoted by the Geological Survey of Italy) and with the study of geological heritage.

Part I
Keynote

Using Water Chemistry to Understand Ancient Maya Land and Water Use Interactions

Sheryl Luzzadder-Beach and Timothy Beach

Abstract
A major goal of Geoarchaeology is to understand complex societies and their relationship with the environment. The Ancient Maya encountered numerous environmental challenges to live in the tropical landscape of Mesoamerica. This research was intended to provide insight into Ancient Maya Collapse and Resilience, and provide environmental models for society now and in the future. Here we focused on using geochemistry, with special attention to water, as a tool for geoarchaeological understanding of ancient Maya land and water use potentials.

Keywords
Geoarchaeology • Geochemistry • Water resources
Mesoamerica • Ancient Maya

1 Introduction

This paper synthesized and presented findings on ancient Maya land and water management in perennial wetlands of the Maya Lowlands (Beach et al. 2015a, b; Luzzadder-Beach et al. 2012; Luzzadder-Beach and Beach 2008, 2009). We presented prior background, then focused on regional studies of water chemistry in the Three Rivers region of Northwestern Belize, comparing surface and ground water results, and wetland results. These watersheds were studied in the context of Ancient Maya land and water use, including wetland field sites we have studied in Northwestern Belize. The focus of our geoarchaeological investigations was centered on the Programme for Belize Rio Bravo Conservation and Management Area, and in the Maya Research Program Blue Creek Study Area (Beach et al. 2015a; Luzzadder-Beach and Beach 2009).

It is also becoming clearer that wetland agriculture was far more extensive than previously known in Northwestern Belize, calling for more mapping, exploration, and ground validation of remote sensing results. In addition to our geochemical investigations, we have taken up this geospatial challenge of testing hypotheses about the extent of Ancient Maya agriculture and other land and water use impacts on the environment through a campaign of airborne LiDAR mapping of the region beginning in 2016.

2 Settings and Methods

The geographic setting of this research is in Northwestern Belize, Central America. Between 1993 and 2018 we have conducted hydrologic and geomorphic studies of the Three Rivers Region of Belize, to understand ancient Maya land and water use (Beach et al. 2015a, b; Luzzadder-Beach et al. 2012; Luzzadder-Beach and Beach 2008, 2009). The Three Rivers include the Rio Bravo, the Rio Azul, and the Booth's River watersheds, draining the La Lucha and Rio Bravo escarpments. This region is on the normally faulted limestone platform of the southern Yucatan Peninsula. In the course of our field work from 2000 to 2018, we have excavated scores of trenches, and collected hundreds of soil and water chemistry samples thus far. This presentation will focus on synthesizing the hydrologic and water chemistry portions of the fieldwork (Beach et al. 2015a, b; Luzzadder-Beach et al. 2012; Luzzadder-Beach and Beach 2008, 2009) and their relationship to other geoarchaeological results for the area.

Our analytical methods have been determining parallel lines of proxy evidence at multiple spatial scales. These methods include excavation to understand soil, geomorphological, and archaeological stratigraphy. We have also engaged in over 2 decades of water quality monitoring to establish the regional waters' geochemical background and links to geologic parent materials. Other lines of evidence

S. Luzzadder-Beach (✉) · T. Beach
The University of Texas at Austin, Austin, TX 78712, USA
e-mail: slbeach@austin.utexas.edu

H. Chenchouni et al. (eds.), *Exploring the Nexus of Geoecology, Geography, Geoarcheology and Geotourism: Advances and Applications for Sustainable Development in Environmental Sciences and Agroforestry Research*, Advances in Science, Technology & Innovation, https://doi.org/10.1007/978-3-030-01683-8_1

include sampling artifacts and ecofacts that indicate past environments and land use adaptations (e.g., soil chemistry and physics, water chemistry, pollen, phytoliths, carbon dating of samples, micromorphology, and macrobotanicals). The spatial extent of these findings was measured using surface and aerial survey, and remote sensing imagery analysis.

3 Results

3.1 Water Quality Results

Upland wetland sites including springs and canals show major geochemical differences from lowland or coastal plain wetland field sites. This result mirrors the water quality differences in the upland versus lowland portions of the Three Rivers watersheds. The results are starkly different in mineral components such as gypsum. This finding supports an argument for anthropogenic origins for the rectilinear wetland field patterns across a broad range of geomorphic and hydrologic conditions in Central America (Beach et al. 2015a, b; Luzzadder-Beach and Beach 2008, 2009). The key result is that SO_4 and Ca are at or near saturation in coastal plain wetland, surface water, and ground water samples, and not so in upland surface and ground water samples (Beach et al. 2015a; Luzzadder-Beach and Beach 2008, 2009). Other cations and anions show similar magnitudes of concentration differences connected with location. We also observed an elevational break point in the gypsum saturation in the Rio Bravo itself, significantly increasing downstream from the Rio Bravo escarpment, indicating a potential change in groundwater chemistry contributions.

4 Discussion

Water chemistry matters in both geomorphic and human contexts. Water chemistry can be a factor in soil and wetland formation processes (Luzzadder-Beach and Beach 2009). Domestic water consumption and use are influenced by water quality, and. water chemistry imposes potential agricultural limitations (Luzzadder-Beach and Beach 2008). The mineral components of water provide a window on underlying geology and hydrologic connectivity. When these factors are considered together at the broadest scale, spatial variation in water chemistry across a region influences the geographies of ecology and human land and water use potentials (Beach et al. 2015a, b; Luzzadder-Beach et al. 2012; Luzzadder-Beach and Beach 2008, 2009).

5 Conclusion

Water chemistry is a major influence on the patterns of land and water use of the ancient Maya of Central America (Beach et al. 2015a, b; Luzzadder-Beach et al. 2012; Luzzadder-Beach and Beach 2008, 2009). While archaeological settlement models of the past have considered amount and accessibility of water as important variables, few have considered the quality of that water as a factor. While it is true that the water in any given river is fleeting, surface water mineral characteristics, and especially the geochemistry of groundwater bodies, provide a window on the long-term influence of underlying geology on water resources. We can therefore consider water quality connected to those mineral inputs or components that change less over human generations, and typically over much longer geologic time frames. These provide insights into the possibilities and limitations of land and water use for ancient human populations (Luzzadder-Beach and Beach 2008), and lessons from the past to guide modern and future societies confronting similar hydrologic challenges, going forward.

References

Beach, T., Luzzadder-Beach, S., Krause, S., Walling, S., Dunning, N., Flood, J., Guderjan, T., Valdez, F.: 'Mayacene' Floodplain and wetland formation in the Rio Bravo watershed of northwestern Belize. The Holocene **25**(10), 1612–1626 (2015a)

Beach, T., Luzzadder-Beach, S., Guderjan, T., Krause, S.: The floating gardens of Chan Cahal: soils, water, and human interactions. Catena **132**, 151–164 (2015b)

Luzzadder-Beach, S., Beach, T.: Water chemistry constraints and possibilities for ancient and contemporary Maya wetlands. J. Ethnobiol. **28**(2), 211–230 (2008)

Luzzadder-Beach, S., Beach, T.: Arising from the wetlands: mechanisms and chronology of landscape aggradation in the northern coastal plain of Belize. Ann. Assoc. Am. Geogr. **99**(1), 1–26 (2009)

Luzzadder-Beach, S., Beach, T., Dunning, N.: Wetland fields as mirrors of drought and the Maya abandonment. Proc. Nat. Acad. Sci. **109**, 3646–3651 (2012)

Part II

Environmental Assessment and Monitoring of Agrisilvicultural Systems

Digital Diffusion for Inclusive Agroecosystems

Chandrashekhar Biradar, Jacques Wery, Fabian Löw, Khaled El-Shama, Rajkumar Singh, Layal Atassi, Jalal Omari, Atef Swelam, Ashutosh Sarkar, Mounir Louhaichi, Boubaker Dhehibi, Enrico Bonaiuti, Bao Le, Theib Oweis, Yashpal Saharawat, Abdoul Aziz Niane, Ahmad Amri, Karan Nadagoudar, Jawoo Koo, Xiangming Xiao, Hrishikesh Ballal, and M. H. Mehta

Abstract

The technological advancements in agriculture have resulted in higher yields but lower ecological efficiency and nutritional value. Little innovations in later sectors such as integrating ecological functions in the production systems have crippled our agro-ecosystems to meet the ever-growing demands. The digitization of the agro-ecosystems has become the most essential entry point for any large scale sustainable developmental entities whether it is, crop diversification, sustainable intensification, input use efficiency, agronomic practices, to restoring ecosystem services and risk management. Recent advances in geoinformatics technology and big-data analytics enabled the diffusion of ecological functions in farm production to achieve the desired return (production follows functions). The overarching goal of the ongoing effort was to build an integrated farming system by leveraging technological diffusion with sound ecological functions to design an 'inclusive agro-ecosystem' for sustainable development. Meta-analytics of farming systems dynamics in spatial domains help quantifying changes, trajectories and drivers under changing climate, demography and degradation process to target site specific developmental interventions and scaling the proven technologies, such as intensification of food legumes in rice fallows, adoption of conservation agriculture, quantification of yield gaps, land/water productivity and transboundary cooperation.

Keywords

Inclusive agroecosystems • Digital diffusion • Ecological intensification • Big-data • Scaling

1 Introduction

Dry lands host an enormous variety of biophysical environments with extremely contrasting socio-economic and demographic conditions. This complexity leads to a wide variability in agricultural functions and productivity across spatio-temporal scales. Therefore, there is a definite need for an ecosystem-based approach for better managing natural resources to improve productivity in a sustainable way that integrates three main goals—environmental health, economic profitability, and social equity. Enormous efforts are underway to gather data and information on agricultural production and related aspects at various scales (http://bigdata.cgiar.org/). There are hundreds of data, tools, apps

C. Biradar (✉) · J. Wery · F. Löw · K. El-Shama · L. Atassi · J. Omari · A. Swelam
International Center for Agricultural Research in Dry Areas (ICARDA), Cairo, Egypt
e-mail: c.biradar@cgiar.org

R. Singh · A. Sarkar
ICARDA, New Delhi, India

A. A. Niane
ICARDA, Beirut, Lebanon

A. Amri
ICARDA, Rabat, Morocco

M. Louhaichi · B. Dhehibi · E. Bonaiuti · B. Le · T. Oweis
ICARDA, Amman, Jordan

Y. Saharawat
ICARDA, Kabul, Afghanistan

K. Nadagoudar
Data Grokr, Bengaluru, India

J. Koo
International Food Policy Research Institute, Washington, DC, USA

X. Xiao
University of Oklahoma, Norman, OK, USA

H. Ballal
Geodesign Hub, Ranelagh, Ireland

M. H. Mehta
Indian Council of Food and Agriculture, New Delhi, India

H. Chenchouni et al. (eds.), *Exploring the Nexus of Geoecology, Geography, Geoarcheology and Geotourism: Advances and Applications for Sustainable Development in Environmental Sciences and Agroforestry Research*, Advances in Science, Technology & Innovation, https://doi.org/10.1007/978-3-030-01683-8_2

and much knowledge available for the increasing productivity but in silos, often without much focus on the ecological functions. There is need for meta-analysis of the best practices with diffusion of ecological functions for designing demand-driven, location-based sustainable agro-ecosystems for resilient livelihood in dry areas.

2 Approach

The dry areas represent a wide agro-biodiversity and the origin of climate resilient crops. However, in the last few decades, the productivity was a major driver rather than functional systems which resulted in the loss of agro-biodiversity in the farming systems. Production follows function and functional agricultural systems are by far more productive and sustainable than dysfunctional systems such as few commodity oriented or mono-cropping systems over long term. This requires dramatic transformational changes from mono systems to multi-cropping systems and re-design of the present agricultural landscape with functional metrics. This entails the systematic quantification of the agricultural production systems and farm analytics at multiple scale(s) with a wide array of data sources to design science-based innovative strategies and principles for inclusive multifunctional agro-ecosystems that are both sustained by nature and sustainable in their nature (Tittonell 2014). The effective use of integrated data in geospatial domains help develop ecological-intensification design eco-zones with location specific crops/varieties, crop sequence, rotation, intensity and crop water productivity (Biradar et al. 2009; Low et al. 2017) under a variable and uncertain climate. Such ecologically balanced agricultural production systems (agro-ecosystems) contribute to the UN's Sustainable Development Goals (SDGs).

3 Use Cases

Several ongoing efforts are made by leveraging the big-data analytics to build use-cases, quantify the agricultural dynamics and understand the variability and drivers for sustainable production systems. Project based site specific studies have been undertaken in different agro-ecologies across the dry areas (http://geoagro.icarda.org/); such as intensification of food legumes in rice fallows in South Asia, adoption of conservation agriculture practices in North Africa, quantification of yield gaps and land degradation in central Asia, trans-boundary land/water productivity at river basins. Here we presented one use-case on 'sustainable intensification of the rice fallows in India as an example and other use-cases can be found in the dedicated portal (http://geoagro.icarda.org/intensification).

3.1 Sustainable Intensification of Rice Fallows

As agricultural production, supply-demand gap continues to rise, the total arable land area is not expected to increase significantly, and future increase of agricultural production will depend more upon sustainable intensification of crop fallows. One such intensification opportunity lies in the use of rice fallows. At present, nearly ~11 million ha rice-fallows are left unproductive. Such large-scale agricultural intensification needs up-to-date information on the dynamics of the rice-based production systems, farm typology, crop and variety specific suitable areas, natural resource base and value chains. Such requirements have led to the development of digital agricultural platform designed to help intensification and crop diversification in crop fallows with food legumes. The digital interactive platform help analyze the farming systems, patterns, dynamics of the crop fallows, length of fallows duration, the start and end of the fallow periods, their seasonality, cropping intensity, etc. helping target sustainable intensification of the legumes crops in the rice fallows with multiple benefits to farmers as well the environment (Fig. 1).

4 Towards Restoring Lost Agro-Ecosystems

Loss of soil health coupled with increasing temperature and water scarcity remain key factors for determining sustainable agricultural productivity and the agri-food systems future. Agriculture production at present relies solely on few crops predominantly under monocultures of industrial agriculture. However, there are still many traditional farming systems with wide array of practices and production functions, which supports the resilience of agro-ecosystems (Altieri et al. 2015; Mehta 2017). The diversified agro-ecosystems is found to be more productive, helps in bridging the yield gaps, more sustainable and resilient to extreme climate events (Astier et al. 2012; Gaba et al. 2015; Low et al. 2018). A regular, accurate and cost effective spatial monitoring of rice-based production systems with satellite data made it feasible to target specific legume intensification. It provides spatially explicit information retrieval about crop productivity, pattern, inter and intra seasonal as well as farm variability which helps develop integrated agro-ecosystems. Such principles need to be embedded in various farm practices and agricultural development strategies, where each one has different effects on productivity, stability and resiliency within the farm system. The digital technology would play key role in an effective diffusion of sustainable agro-ecosystems and targeting a wider and faster adaptation to changing climate and demography. Finally, only by creating policies endorsed diffusion of digital technology that integrates ecological functions with economic and social welfare can we promote the adoption of sustainable agricultural systems across the scales.

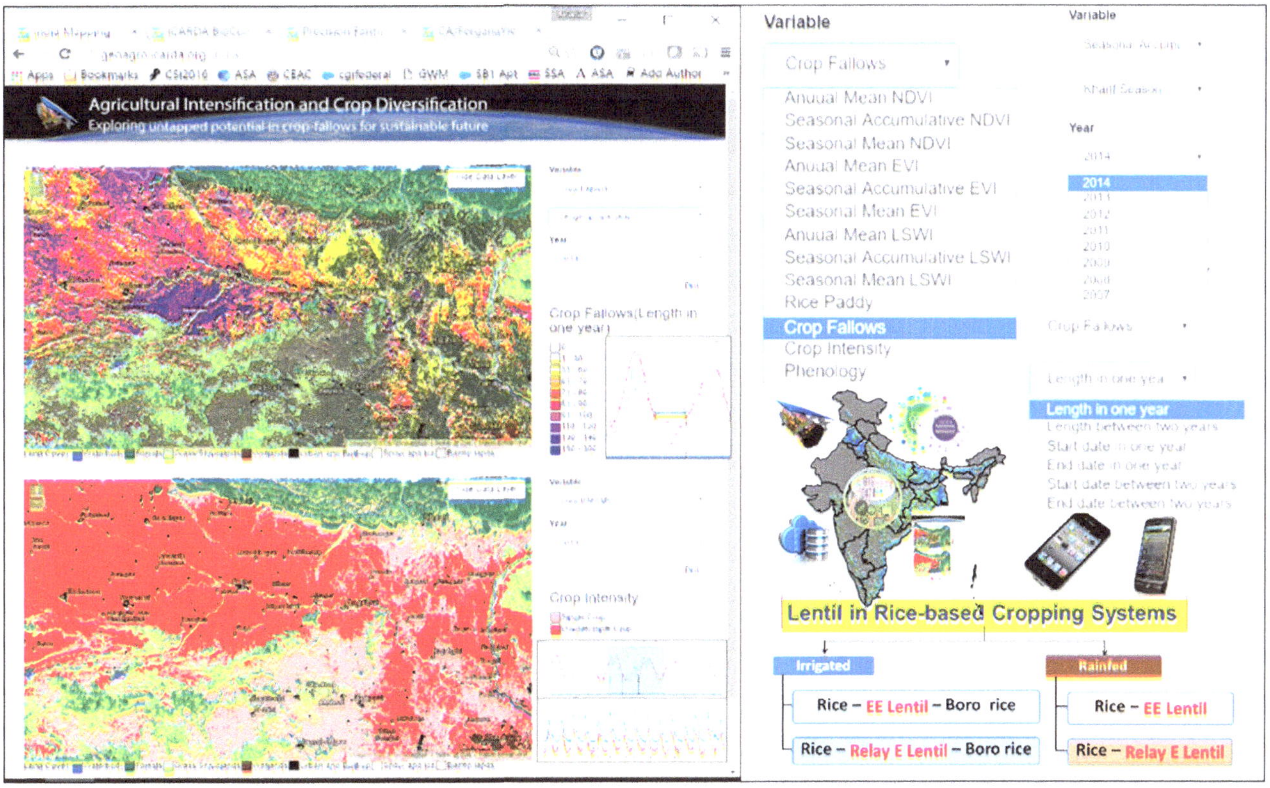

Fig. 1 Spatial Big-data based digital agricultural platform for ecological intensification of rice fallows in India

References

Altieri, M., Nicholls, C., Henao, A., Lana, M.: Agroecology and the design of climate change-resilient farming systems. Agron. Sustain. Dev. **35**, 869–890 (2015)

Astier, M., García, L., Galván, Y., González, E., Masera, R.: Assesing the sustainability of small-farmer natural resource management systems. A critical analysis of the MESMIS program (1995–2010). Ecol. Soc. **17**(3), 25 (2012)

Biradar, C.M., Thenkabail, P.S., Noojipady, P., Yuanjie, L., Dheeravath, V., Velpuri, M., Turral, H., Gumma, M.K., Reddy, O.G.P., Xueliang, L.C., Xiao, X., Schull, M.A., Alankara, R.D., Gunasinghe, S., Mohideen, S.: A global map of rainfed cropland areas (GMRCA) at the end of last millennium using remote sensing. Int. J. Appl. Earth Obs. Geoinf. **11**(2), 114–129 (2009)

Gaba, S., Lescourret, F., Boudsocq, S., Enjalbert, J., Hinsinger, P., Journet, E., Navas, M., Wery, J., Louarn, G., Malézieux, E., Pelzer, E., Prudent, P., Lafontaineet, H.: Multiple cropping systems as drivers for providing multiple ecosystem services: from concepts to design. Agron. Sustain. Dev. **35**(2), 607–623 (2015)

GeoAgro Homepage. http://geoagro.icarda.org/. Accessed 14 May 2018

Low, F., Biradar, C., Fliemanna, E., Lamers, J., Conrad, C.: Assessing gaps in irrigated agricultural productivity through satellite earth observations - A case study of the Fergana Valley, Central Asia. Int. J. Appl. Earth Obs. Geoinf. **59**, 118–134 (2017)

Low, F., Biradar, C., Dubovyk, O., Fliemann, E., Akramkhanov, A., Vallejo, A.N., Waldner, F.: Regional-scale monitoring of cropland intensity and productivity with multi-source satellite image time series. GIScience Remote Sens. **55**(18), 539–567 (2018)

Mehta, M.H.: Eco Agri Revolution: Practical Lessons and The Way Ahead. New India Publishing Agency—Nipa, New Delhi, India (2017)

Tittonell, P.A.: Ecological intensification of agriculture—sustainable by nature. Curr. Opin. Environ. Sustain. **2014**(8), 53–61 (2014). https://doi.org/10.1016/j.cosust.2014.08.006

Climate and Territorial Suitability for the Vineyards Developed Using GIS Techniques

Matteo Gentilucci, Maurizio Barbieri, and Peter Burt

Abstract
This study assessed the wine vocation of two grapevine cultivars (Maceratino and Sangiovese) in Macerata Province (Central Italy) for different periods: 1931–1960, 1961–1990 and 1991–2014. The maps of viti-cultural potential were edited using GIS software, on the basis of late frosts, optimum temperatures, water requirement and slope. Raster math allowed the summation of the parameters considered, also assessing climate change between periods, in relation to the crops. The aim of this research was to provide a valuable tool for more rational spatial planning.

Keywords
Vine • Grapevine • GIS • Climate • Soil

1 Introduction

The viticultural potential of wine regions is influenced by many factors that can be summarized in climate, environment and soil. This study aimed to identify the adaptation of some grapevine varieties (Maceratino and Sangiovese) cultivated in the Province of Macerata (Central Italy, on the Adriatic side), land of wine, through some maps created using GIS software. In order to evaluate the optimal conditions for the growth of the vine, some environmental factors were considered: late frosts, optimum temperature in the growing season, land slope and water requirement. This research purposes the innovative task of being a profitable instrument for wine farmers to plan the most suitable areas to be planted with vines and a tool to evaluate climate change with regard to the plant well-being. There are different topics on this study, but above all the relation between climate (Moriondo and Bindi 2007) and phenology for vines cultivation, is the main one; in fact, phenological phases of fruit trees are closely related to temperatures. The phenological response of plants, as end of dormancy, has been studied since the 1970s with the Utah model (Richardson et al. 1974), until the present day with more accurate models such as that of Cesaraccio et al. (2004). Instead of the other phenological stages, there are different models (Fraga et al. 2016; Zapata et al. 2016). However, a similar model to that of Cesaraccio has been chosen, namely the Gentilucci model (Gentilucci 2017). It was calibrated in Central Italy, with threshold temperatures of growing (warm days and no warm days) without a set date to start the GDD (growing degree days) count, but from the bud swelling of each year. Therefore the outcome of this analysis was to provide a "wine vocation map" using GIS software, that takes into account not only temperature (Huglin 1978; Hall and Jones 2010) but also the other variables mentioned above.

2 Methods

The survey was carried out for 3 different standard periods: 1931–1960, 1961–1990 and 1991–2014. The outcome is represented by the overlap of some layers, which leads to the creation of a wine vocation map for each considered period. Four factors have been considered: late frost hazard, optimum temperature, water requirement and land slope. Frost hazard maps calculate the probability of having temperatures below zero during the growing season (harmful for the plant), using the probability kriging though GIS software (ArcGis). Concerning optimal temperatures for the

M. Gentilucci (✉)
University of Camerino, 62032 Camerino, Italy
e-mail: matteo.gentilucci@unicam.it

M. Barbieri
Sapienza University of Rome, 3029 Rome, Italy

P. Burt
University of Greenwich, ME4 4TB Chatham Maritime, Kent, UK

H. Chenchouni et al. (eds.), *Exploring the Nexus of Geoecology, Geography, Geoarcheology and Geotourism: Advances and Applications for Sustainable Development in Environmental Sciences and Agroforestry Research*, Advances in Science, Technology & Innovation, https://doi.org/10.1007/978-3-030-01683-8_3

studied vines, they are known from a previous research (Gentilucci 2017) and were used as a limit for the probability cokriging and summarized in a single map for each period, through the raster math tool. To estimate the plant water requirement, reference evapotranspiration was calculated through the Hargreaves-Samani formula (Hargreaves and Samani 1982). The evapotraspiration of all the weather stations and for all the periods was interpolated through geo-statistical methods (ordinary kriging); the outcome was multiplied by Kc, the single crop coefficient (Doorenbos and Kassam 1979), in order to obtain the crop evapotranspiration (Allen et al. 1998). Subsequently the effective rainfall was estimated (Doorenbos and Pruit 1977) to obtain the spatial distribution of available water capacity (AWC), that is the maximum amount of water available for grapevines. Furthermore, the extractable soil water (RU) in relation to the rooting depth of grapevine was investigated, multiplying AWC by the rooting depth. In order to obtain the water balance for grapevines, groundwater was investigated through some parameters like extractable soil water, effective rainfall and crop evapotranspiration for grapevines.

Groundwater = RU (to the starting point) + effective rainfall + ETc

The groundwater was estimated in percentage compared to the maximum amount of the extractable soil water balance down to a depth of 80 cm. According to previous studies, it is possible to assess [ERSA] the amount of optimum extractable soil water for each month with a value ranging from 0 to 1. The proposed method takes into account both water stress and water excess, to evaluate the optimum conditions for the grapevines. In fact, in August and September, water stress can be considered an advantage, above all for the quality of wine grape, while in the other months there is a problem for the grapes growing. Finally, the values obtained with this procedure were averaged between all the months of the growth interval for each reference period (1931–1960, 1961–1990 and 1991–2014). Finally, the slope angle, which can be a limiting factor for vines, was studied because in high slope values, soil erosion leads to nutrient losses. A lack of slope, however, can cause roots rottenness or sugar content decrease in berries.

3 Results

The above process resulted in four maps for each period (1931–1960, 1961–1990 and 1991–2014): a frost hazard map, optimum temperatures map, water requirement map, and a slope map. All the maps were standardized and for each investigated period an average of the four previously described maps was calculated. The result of this computation is another map that has a coefficient between 0 (minimum adaptation of the grapevine to the environment) and 1 (maximum adaptation of the grapevine to the environment) (Fig. 1). This procedure identified the most suitable areas for the grapevine related to the investigated varieties.

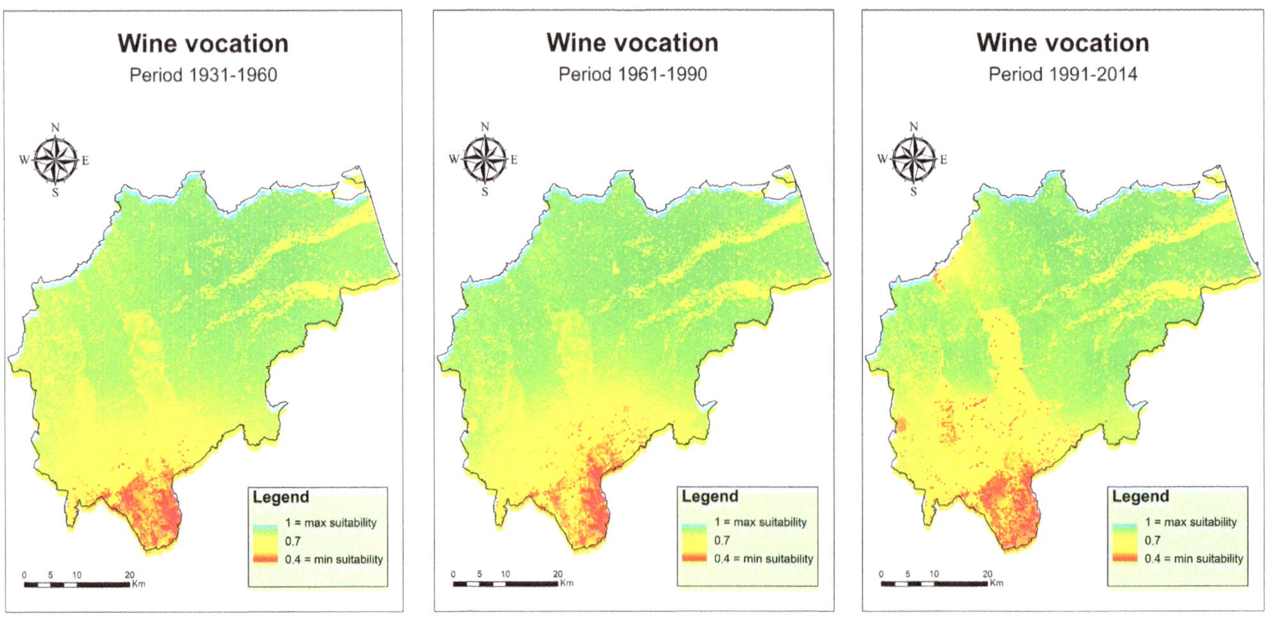

Fig. 1 Wine vocation maps for the Province of Macerata for the 3 periods (1931–1960, 1961–1990 and 1991–2014)

4 Discussion

The analysis estimates at best the distribution of wine vocation in the Province of Macerata, allowing a quick visualization of the best and the worst areas. The worst areas for vineyards cultivation are located in the mountains, piedmont and river-bed zones. Therefore the software characterization is in line with the expected results and corresponds to the areas of high viti-cultural concentration. In fact the resulting maps fit quite well with the territorial identification of the DOC ("Denominazione di Origine Controllata", i.e. a brand of quality for wines). In fact, almost all the green areas are located in the identified zone for the DOC-wine production such as: Verdicchio di Matelica, I terreni di San Severino, Colli maceratesi, San Ginesio and Serrapetrona. It is also interesting to observe how climate change, represented by an increase in average temperatures from the past (1931–1960) to the present (1991–2014), generates a reduction, with the same trend, of the optimal areas for vine cultivation (Fig. 1).

5 Conclusion

Although the weight of the four parameters has not been diversified because of the lack of experimental data, the differentiation displayed in the map (Fig. 1) can be a preliminary investigation. This study could become an operational tool for farmers to consciously choose the location of their own business, assessing its potential profitability. Further developments would be desirable in the future by applying this procedure to other crops in order to assess the influence of climate change in relation to plant adaptation.

References

Allen, R.G., Pereira, R., Raes, D., Smith, M.: Crop evapotranspiration: guidelines for computing water requirements. FAO irrigation and drainage paper no. 56, Food and Agricultural Organization, Rome, Italy (1998)

Cesaraccio, C., Spano, D., Snyder, R.L., Duce, P.: Chilling and forcing model to predict budburst of crop and forest species. Agric. For. Meteorol. **126**, 1–13 (2004)

Doorenbos, J., Kassam, A.H.: Yield response to water. FAO irrigation and drainage paper no. 33. FAO, Rome, Italy (1979)

Doorenbos, J., Pruit, W.O.: Guidelines for predicting water requirements. FAO irrigation and drainage paper 24. FAO, Rome, Italy (1977)

ERSA (Agenzia regionale per lo sviluppo rurale), Gestione dell'irrigazione: http://www.ersa.fvg.it/tematiche/suoli-e-carte-derivate/cartografia-derivata/zonazione-viticola/zona-a-d-o-c-collio/manuale-duso-del-territorio/gestionedellirrigazione/?searchterm=gestione%20dell%27irrigazione. Last accessed 27 Apr 2018

Fraga, H., Santos, J., Moutinho-Pereira, J., Carlos, C., Silvestre, J., Eiras-Dias, J., Malheiro, A.: Statistical modelling of grapevine phenology in Portuguese wine regions: observed trends and climate change projections. J. Agri Sci **154**(5), 795–811 (2016)

Gentilucci, M.: Grapevine prediction of end of flowering date. In: Acts of the 1st Euro-Mediterranean Conference for Environmental Integration (EMCEI), pp. 1231–1234, Springer, Tunisia (2017)

Hall, A., Jones, G.: Spatial analysis of climate in winegrape-growing regions in Australia. Aust. J. Grape Wine Res. **16**, 389–404 (2010)

Hargreaves, G.H., Samani, Z.A.: Estimating potential evapotranspiration. J. Irrig Drainage Eng, ASCE **108**(IR3), 223–230 (1982)

Huglin, P.: Nouveau Mode d'Évaluation des Possibilités Héliothermiques d'un Milieu Viticole. Comptes Rendus de l'Académie d'Agriculture de France **64**, 1117–1126 (1978)

Moriondo, M., Bindi, M.: Impact of climate change on the phenology of typical mediterranean crops. Ital. J. Agrometeorol. 5–12 (2007)

Richardson, E.A., Seeley, S.D., Walker, D.R.: A model for estimating the completation of rest for Redhaven and Elberta peach trees. HortScience **9**(4), 331–332 (1974)

Zapata, D., Salazar-Gutierrez, M., Chaves, B., Keller, M., Hoogenboom, G.: Predicting key phenological stages for 17 grapevine cultivars (*Vitisvinifera* L.). Am. J. Enol. Viticulture (2016)

Projected Small-Scale Range Reductions of *Cedrus atlantica* Forests Due to Climate Change at the Belezma National Park (Algeria)

Abdelkrim Arar, Slimane Tabet, Yacine Nouidjem, Rabah Bounar, and Haroun Chenchouni

Abstract

The aim of this study was to highlight the impact of climate change on the spatial distribution of *Cedrus atlantica* forests in the Belezma National Park (BNP) in Algeria. Maximum entropy (MaxEnt) species distribution modeling approach combined with 19 climatic, 15 soil-related and 15 topographical factors were implemented here for the spatial distribution of *Cedrus atlantica*. The future predictions of both horizons 2050 and 2070 (CCSM4 model) were carried out to understand the future suitable range of the species. Results showed that the model MaxEnt was efficient and revealed significantly higher measurement of the AUC value (Area Under Curve) compared to the random prediction. AUC was 0.986 for training and 0.980 for the test data. The response curves of species occurrence *vs.* environmental variables showed that climatic variables have the most significant contributions compared to edaphic and topographic factors. A very high decrease in suitable area of *C. atlantica* was forecast on the basis of our predictions for 2070. This indicates that the species has a little chance to survive under the conditions of the future climate change scenarios. Our findings incite us to call for urgent actions in order to reduce the vulnerability of this species' habitats in front of climate change.

Keywords

Climate change • *Cedrus atlantica* • Ecological niche North African forests • Species distribution modeling MaxEnt • Belezma national park

1 Introduction

Climate change is one of the topical issues that worries man. Recent studies on the basis of prediction modeling approaches showed that due to the increase in greenhouse gas (GHG) concentrations, the global temperature will increase in the future (Solomon et al. 2009). The ecological modeling allows the description of natural phenomena and their spatial and temporal dynamics through differential equations (Phillips et al. 2006; Elith et al. 2011). Species Distribution Models (SDM) are crucial tools in ecological sciences as they provide a better understanding of species ecology (Tabet et al. 2018). These models include MaxEnt (Maximum Entropy), which is a model based on automated statistical techniques (Phillips et al. 2006). Several other modeling approaches can be used to develop SDMs, including BIOCLIM, GAMs, GARP, ANN etc. (Elith and Graham 2009).

The Atlas cedar (*Cedrus atlantica* (Endl.) Manetti ex Carrière) is an endemic species to North Africa. Quite widespread in Morocco (Rhanem 2011), *Cedrus atlantica* is a rare species in Algeria and forms forest formations occupying high altitudes. Knowing its particular ecological requirements in terms of soil moisture and low temperature, the severe droughts of the previous years and climatic changes are the cause of this species' high disappearance rate in several regions in North Africa (Demarteau et al. 2007; Chenchouni et al. 2008; Slimani et al. 2014; Alba-Sánchez et al. 2015; Cheddadi et al. 2017; Abel-Schaad et al. 2018).

This study aimed to evaluate the climate change effects on the spatial distribution of *Cedrus atlantica* at the Belezma National Park—Algeria (BNP), by analyzing the species

A. Arar · Y. Nouidjem · R. Bounar
Department of Life and Nature Science, Faculty of Life and Nature Sciences, Mohamed Boudiaf University, 28000 M'Sila, Algeria
e-mail: ararabdelkrim@gmail.com

S. Tabet
Abdlahfid Boussouf University Center, 43012 Mila, Algeria

H. Chenchouni (✉)
Department of Natural and Life Sciences, FSESNV, University of Tebessa, 12002 Tebessa, Algeria
e-mail: chenchouni@gmail.com

H. Chenchouni
Laboratory of Natural Resources and Management of Sensitive Environments 'RNAMS', University of Oum-El-Bouaghi, 04000 Oum-El-Bouaghi, Algeria

H. Chenchouni et al. (eds.), *Exploring the Nexus of Geoecology, Geography, Geoarcheology and Geotourism: Advances and Applications for Sustainable Development in Environmental Sciences and Agroforestry Research*, Advances in Science, Technology & Innovation, https://doi.org/10.1007/978-3-030-01683-8_4

distribution responses to climatic, edaphic and topographic factors, and identify suitable future habitats for this species under various climate scenarios for the 2050 and 2070 projections based on the CCSM4.0 climate model.

2 Materials and Methods

The current study was carried out at the Wilaya 'Province' of Batna in northeastern Algeria. The Belezma National Park (BNP) (longitude: 5.89E–6.32E, latitude: 35.51N–35.70N) is located at the intersection zone between the two Tellian and Saharan mountain chains of Algeria (Fig. 1) and stretches over an area of 26,250 ha with a position at the western end of the Aurès Massifs (Sadine et al. 2012; Slimani et al. 2014).

MaxEnt is a general modeling method based on the best approximation of the distribution of an unknown probability and therefore this distribution must have a maximum entropy (Phillips et al. 2006). The choice of this method was based on the important peculiarities it has compared to other models, among which, the method is adapted to use presence data only in addition to the possibility of using many interacting variables. From the observed points of presence, MaxEnt estimates the most consistent potential distribution of the species relative to the predicted value for each environmental variable (Phillips et al. 2006). The high predictive performances of MaxEnt have been confirmed in several studies (Elith and Graham 2009; Elith et al. 2011). The other models of regression types like GLM, and GAM, among others, are limited in ecology to correlative type applications for species occurrence predictions (Elith et al. 2011).

This study used the presence data of *Cedrus atlantica* for modeling its spatial distribution at BNP under the SDM model 'MaxEnt'. The database was validated using field surveys that served to collect species' presence points using a GPS (Fig. 1). To ensure realism and robustness of the results of the spatial distribution of the target species, a total of 49 environmental variables were used (https://doi.org/10.6084/m9.figshare.6207374). These data were obtained from various sources and included 19 climatic variables (www.worldclim.org), 15 edaphic variables (https://soilgrids.org), and 15 variables related to topography (https://gdex.cr.usgs.gov/gdex). For future climate data we used two RCP scenarios (RCP 4.5 and 8.5) due to the low variation of climatic parameters between all scenarios in the study area. These two scenarios consist of the moderate and high emissions and concentration scenarios with GHG radiative forcing reaching 4.5 W m^{-2} and 8.5 W m^{-2}, respectively.

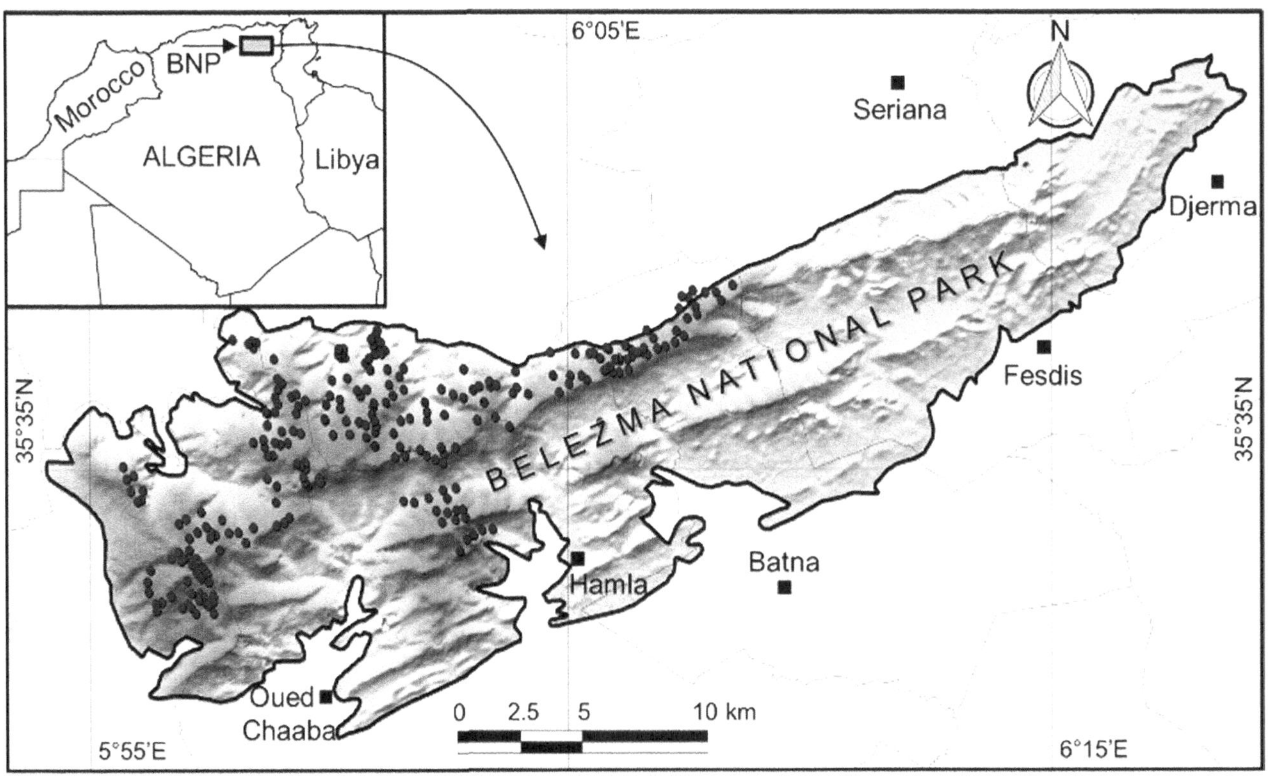

Fig. 1 Location map of the Belezma National Park (northeastern Algeria). The scattered grey circles indicate the validated presence points of *Cedrus atlantica* stands

3 Results

3.1 Predicted Current Species Distribution

The habitat suitability distribution map of *C. atlantica* was developed. It showed probabilities of the suitable distributional habitats that were estimated to an area of 2600 ha at BNP (Fig. 2a). The MaxEnt model identified a list of ecological factors that have the greatest contribution to the current distribution of the species at BNP. The response curves show the responses of the species to gradient variations of the most contributing ecological factors (Fig. 3). MaxEnt receiver operating characteristic (ROC) curve of sensitivity versus cedar specificity curve showed that the obtained evaluation results confirm the high performance of the model with an AUC = 0.986 for the training data and an AUC = 0.980 for the test data, which were both significantly higher that the random prediction (AUC = 0.5) (Fig. 2b, c). This means that

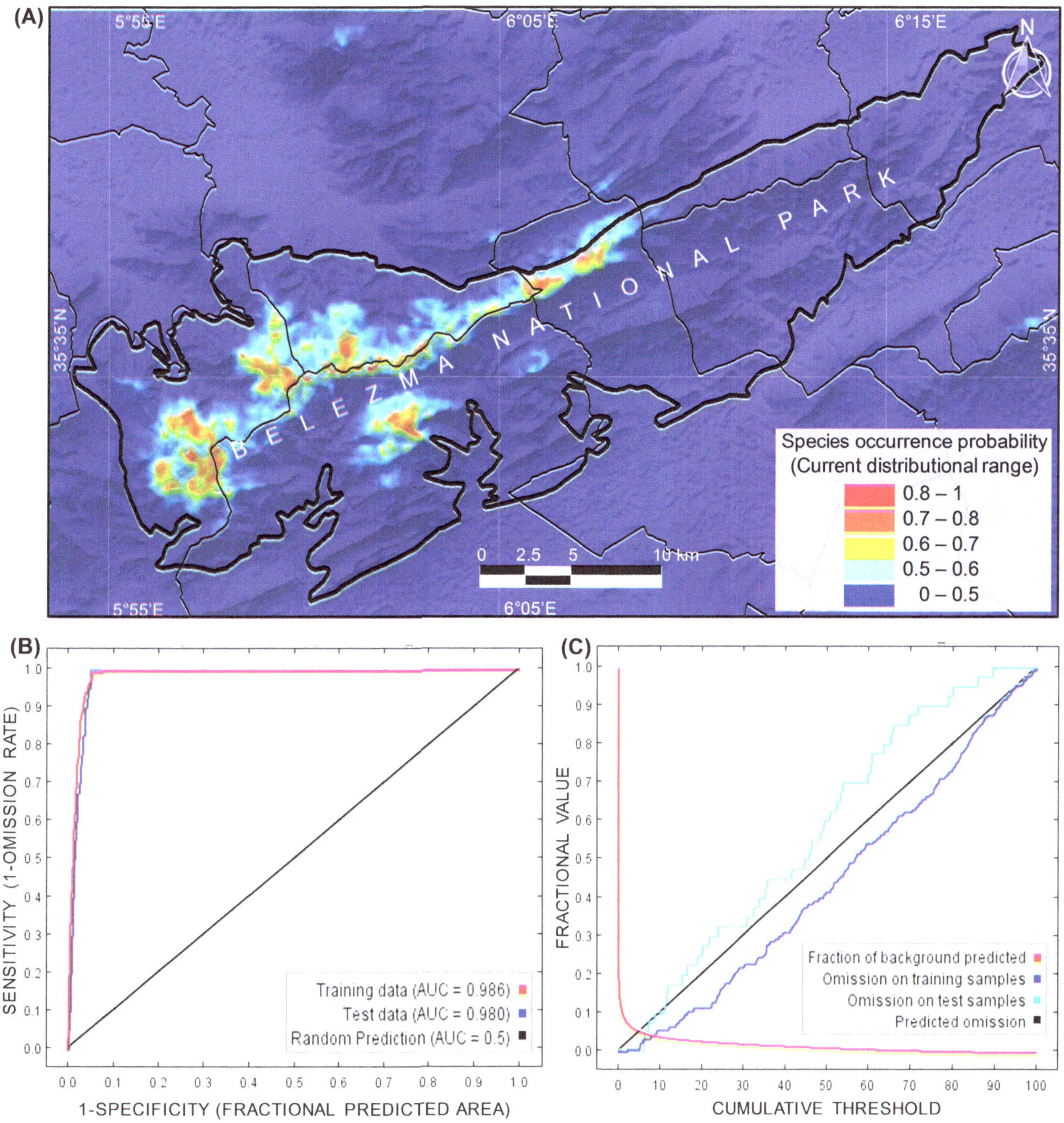

Fig. 2 **a** Map of the probability of presence of the atlas cedar (*Cedrus atlantica*) for the current period at The Belezma National Park (NE Algeria), **b**. MaxEnt receiver operating characteristic (ROC) curve indicating AUC of training, test and random prediction data, **c** omission rate of tested and predicted *Cedrus atlantica* suitable areas

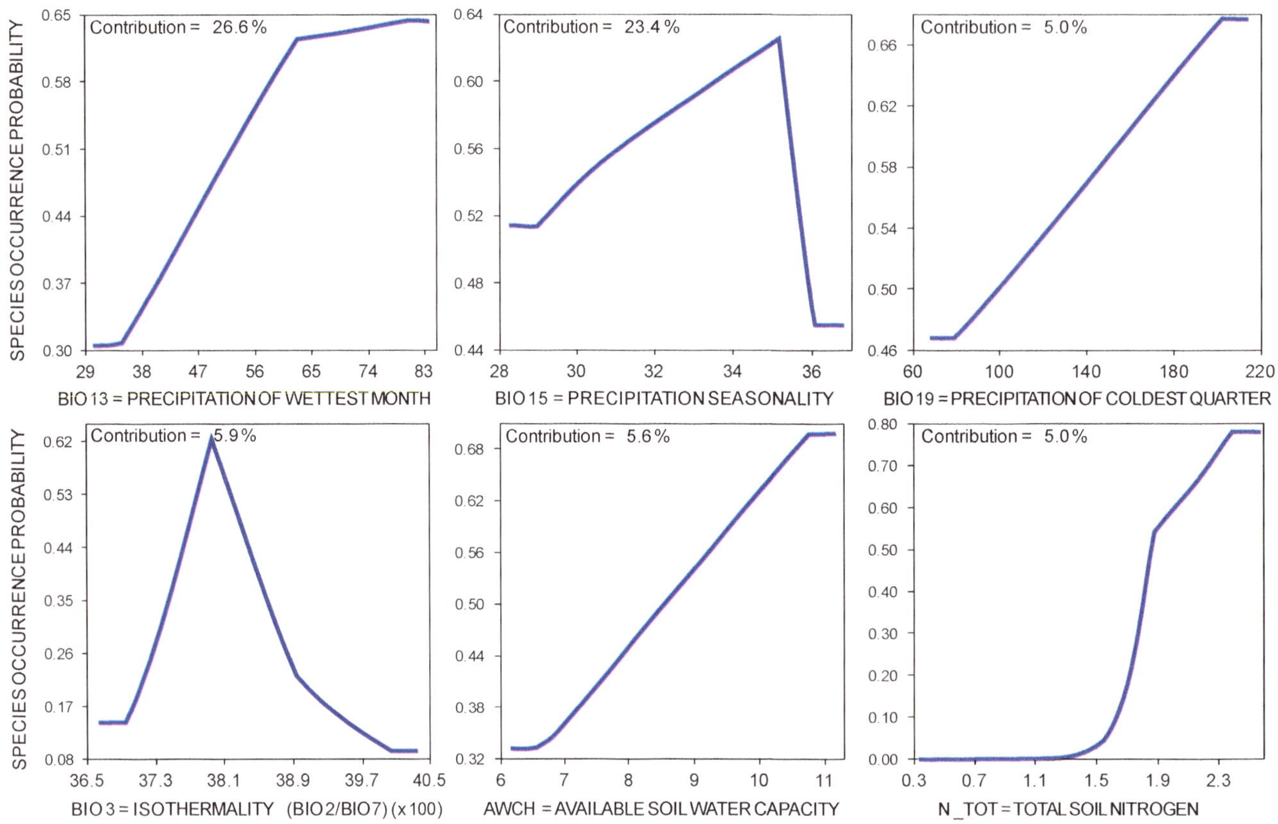

Fig. 3 Response curves of five environmental factors (data 1950–2000) with the highest contributions (in %) on the current distribution of *Cedrus atlantica* in the Belezma National Park, Algeria

the model performed very well against the random model. Our results showed the great influence of climatic variables on the distribution of *Cedrus atlantica*, compared to edaphic and topographic variables that have little contribution (https://doi.org/10.6084/m9.figshare.6207374).

3.2 Predicted Future Species Distributional Range

The results of the predicted Atlas cedar ecological niche showed the complete absence of this species in all the scenarios (RCP4.5 and RCP8.5) of the two future horizons 2050 and 2070. This indicates that the projected future climatic conditions will be undeniably unfavorable for the occurrence of the Atlas cedar in the BNP and eventually for the rest of cedar forests in northeastern Algeria.

4 Discussion

Future predictions of *Cedrus atlantica* distribution indicate that the suitable habitat areas decrease very significantly for both the 2050 and 2070 projections. These results confirm the species' sensitivity to seasonal and inter-annual temperature and precipitation variations (Chenchouni et al. 2008; Allen et al. 2010); which triggered the recent reductions of the species range that shifted toward high altitudes, which constitute refuge areas to the species (Cheddadi et al. 2017). The same species' range shrinkages were also observed in Moroccan *Cedrus atlantica* forests (Demarteau et al. 2007; Abel-Schaad et al. 2018). This drought-intolerance results in a significant and alarming decrease (up to 100%) in the area of suitable habitats for the development Atlas cedar by 2070 for both RCP4.5 and RCP8.5 scenarios, which represent the moderate and extremely high pessimistic predictions of climate change, respectively. This means that the challenge of conservation of the species is major in terms of climate change. Such an alarming result is confirmed by the intense decline and mass mortality of trees of this species at BNP over the last two decades (Chenchouni et al. 2008). It is noteworthy mentioning that prediction results obtained by the modeling technique are, in fact, somewhat exaggerated because the model did not take into account the physiological characteristics and adaptive capacities of the species. This is actually a disadvantage of MaxEnt, which biased a good knowledge of species' responses to current and future environmental variations.

Added to this, is the lack of data on the influences of socio-economic activities and the use of small-scale geographic data.

5 Conclusion

The current work showed that the forest stands of BNP, where the Atlas cedar is the main species, have shown a spatiotemporal change in their distribution, which is a natural response to climate change. The sharp decrease may lead to a complete extinction of this species during future horizons of 2050–2070 under the influence of climate changes which are the main contributing factors with their direct impacts on the distribution of *Cedrus atlantica*. This study was initially carried out to in order to estimate and evaluate the magnitude of such changes over time and space. Highlighting the effects of the environmental gradients on the current and future distribution of *Cedrus atlantica* forests is very useful for the management and decision-making in terms of conservation of this important species and the accompanying biodiversity.

References

Abel-Schaad, D., Iriarte, E., López-Sáez, et al.: Are *Cedrus atlantica* forests in the Rif Mountains of Morocco heading towards local extinction?. The Holocene, 0959683617752842 (2018)

Alba-Sánchez, F., López-Sáez, J.A., Nieto-Lugilde, D.: Long-term climate forcings to assess vulnerability in North Africa dry argan woodlands. Appl. Veg. Sci. **18**(2), 283–296 (2015)

Allen, C.D., Macalady, A.K., Chenchouni, H., et al.: a global overview of drought and heat-induced tree mortality reveals emerging climate change risks for forests. For. Ecol. Manage. **259**(4), 660–684 (2010)

Cheddadi, R., Henrot, A.J., François, L., et al.: Microrefugia, climate change, and conservation of *Cedrus atlantica* in the Rif mountains, Morocco. Frontiers in Ecology and Evolution **5**, 114 (2017)

Chenchouni, H., et al.: The deterioration of the Atlas Cedar (*Cedrus atlantica*) in Algeria. Int. Conf. "Adaptation of forests and forest management to changing climate with emphasis on forest health", FAO/IUFRO, Umeå, Sweden, p. 51 (2008)

Demarteau, M., Francois, L., Cheddadi, R., Roche, E.: Réponses de *Cedrus atlantica* aux changements climatiques passés et futurs. Geo-Eco-Trop **31**, 105–146 (2007)

Elith, J., Graham, C.H.: Do they? How do they? why do they differ? On finding reasons for differing performances of species distribution models. Ecography **32**(1), 66–77 (2009)

Elith, J., Phillips, S.J., Hastie, T., Dudík, M., Chee, Y.E., Yates, C.J.: A statistical explanation of MaxEnt for ecologists. Divers. Distrib. **17** (1), 43–57 (2011)

Phillips, S.J., Anderson, R.P., Schapired, R.E.: Maximum entropy modeling of species geographic distributions. Ecol. Model. **190**, 231–259 (2006)

Rhanem, M.: Aridification du climat régional et remontée de la limite inférieure du cèdre de l'Atlas (*Cedrus atlantica* Manetti) aux confins de la plaine de Midelt (Maroc). Physio-Géo **5**, 143–165 (2011)

Sadine, S.E., Alioua, Y., Chenchouni, H.: First data on scorpion diversity and ecological distribution in the National Park of Belezma, northeast Algeria. Serket **13**, 27–37 (2012)

Slimani, S., Derridj, A., Gutierrez, E.: Ecological response of *Cedrus atlantica* to climate variability in the Massif of Guetiane (Algeria). Forest Systems **23**(3), 448–460 (2014)

Solomon, S., Plattner, G.K., Knutti, R., Friedlingstein, P.: Irreversible climate change due to carbon dioxide emissions. Proc. Natl. Acad. Sci. **106**(6), 1704–1709 (2009)

Tabet, S., Belhemra, M., Francois, L., Arar, A.: Evaluation by prediction of the natural range shrinkage of *Quercus ilex* L. in eastern Algeria. Forestist 68(1), 7–15 (2018)

Sinuosity of Meandering Channels in Upper Krishna River Basin, India

Suchitra Pardeshi, Sudhakar Pardeshi, and Pallavi Kulkarni

Abstract

Krishna River is one of the important drainage systems in South India. Krishna River and its tributaries, like Panchganaga and Dudhaganaga Rivers, show a meandering planform morphology. The Krishna River and its tributaries originate in the high rainfall region of Western Ghat. Flood discharge conditions of these channels cause change in channel morphology like the formation of chute channels or cut-off processes. To understand the cut-off channel formation, hydraulic and topographic sinuosity of channel segments were calculated. To this end, three segments of each channel were considered first, before meander, second at the meander and third after the meander of the channel were calculated. It is observed that when the hydraulic sinuosity is greater than the topographical sinuosity, chute channels are formed in the upper Krishna Basin.

Keywords

Krishna River • Sinuosity index • Chute channel Meandering channels • Western Ghat

1 Introduction

The significance of streams in the evolution of landscapes has long fascinated geomorphologists, hydrologists and geologists (Muller 1968). Knowledge of the structure, energy and dynamics of a river is important to understand the characteristics of a river to manage and protect it (Ebisemiju 1994). When rivers flow from the source through different landscapes, they attain different patterns. According to Leopold and Wolman (1957) the channel pattern is traditionally classified as straight, meandering and braiding types. Schumm and Khan (1972) explained sinuosity as the degree to which a river departs from a straight line. The distance between two points of the stream measured along the channel and divided by the straight line distance between the same two points is called the sinuosity ratio (Brice 1984; Ebisemiju 1994). Sinuosity is basically of two types; one is topographic and the other is hydraulic. When a stream departs from a straight line path due to topographic control is called topographic sinuosity. Hydraulic sinuosity is caused when the stream path deviates from straight line path due to hydraulic factors such as high discharge. Muller (1968) mentioned that in topographic sinuosity the stream adjusts with surface and in hydraulic sinuosity flow of stream attains sinuosity.

The present study was carried out to understand the sinuosity in Krishna River through different sinuosity indices.

The Krishna River and its tributaries show very high sinuosity (see Fig. 1).

S. Pardeshi
Prof. Ramkrishna More ACS College, Akurdi, Pune, 411044, India
e-mail: suchitrapardeshi@gmail.com

S. Pardeshi (✉) · P. Kulkarni
Department of Geography, Savitribai Phule Pune University, Pune, 411007, India
e-mail: sdpardeshi@rediffmail.com

2 Materials and Methods

The information about the study area was collected from sources like Government of India (2014) and Survey of Indian topographic maps. Besides, theodolite surveys were carried out. Based on field observation, the morphological data were generated to understand the topography, channel morphology and processes.

Morphological data (e.g. slope, elevation) were extracted from SOI topographic maps (47 L/10, 1973) and ASTER-DEM. The Google earth was also used to create the thematic data and each data were digitized.

For the sinuosity index calculation, three different segments from each channel were selected and for each segment the distances between two points of meandering path and

H. Chenchouni et al. (eds.), *Exploring the Nexus of Geoecology, Geography, Geoarcheology and Geotourism: Advances and Applications for Sustainable Development in Environmental Sciences and Agroforestry Research*, Advances in Science, Technology & Innovation, https://doi.org/10.1007/978-3-030-01683-8_5

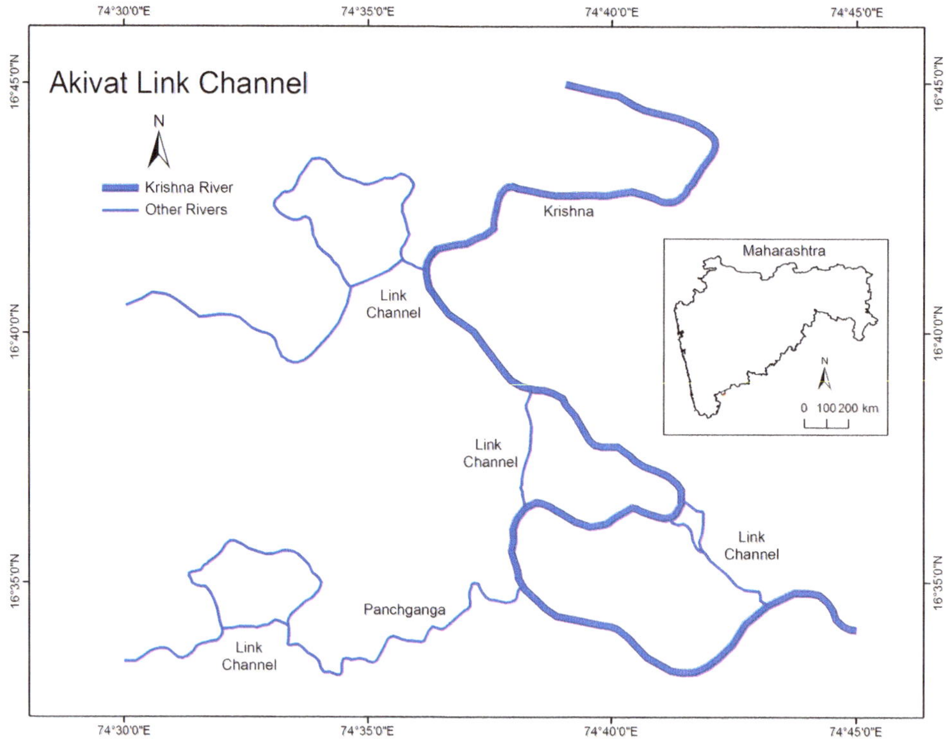

Fig. 1 Location map of study area

straight line distance between those points were measured (Aswathy et al. (2008)). By using the channel index (CI) that is the ratio of channel length to straight line distance and Valley Index (VI) that is the ratio of valley length to straight line distance, the calculation of hydraulic sinuosity index (HSI) and topographic sinuosity index (TSI) was achieved (Muller 1968).

3 Results

3.1 Cut off or Link Channels

It is observed that in the study area, wherever the river is meandering, the main stream is unable to accommodate the flood discharge and water follows a straight and the shortest path; these are called as 'link channels'. Such links are observed in the tributary channels before they meet the main one. In the main channel, links are found in the downstream of the confluences.

The sinuosity index of Krishna River ranges from 1.28 to 4.13 (Table 1). The highest sinuosity is observed near the link channel of Akiwat. Before Akiwat, Krishna River also flows in meandering path but its sinuosity is relatively less and therefore does not form a link channel (Table 1).

Panghganga River is a right bank tributary and also forms its link channel near Kurundwad and the river has more sinuosity index near the link channel (Table 1).

Dudhganga joins Krishna after Akiwat link channel. This river has formed its link channel near Sadalgi where the sinuosity index of this river is high.

Table 1 Sinuosity index (All values are in meters)

Sr. no.	Channel	Location	Channel length	Straight length	Sinuosity index
1	Krishna	Akiwat 1	27041.795	13072.660	2.0685
2		Akiwat 2	13829.874	4418.582	3.1299
3		Akiwat 3	19653.716	4757.945	4.1307
4		Akiwat 4	7620.221	5949.379	1.2808
5		Akiwat 5	71655.666	21569.149	3.3221
6	Dudhganaga	Sadalgi	12281.420	2486.288	4.9396
7	Panchganga	Shirdon 1	25465.877	9919.829	2.5671
8		Kurundwad 2	15322.609	2755.394	5.5609

Table 2 Sinuosity indices for Krishna Basin

Index type	Values	% Values	Formula used
HSI	0.7174	71.74	(CI − VI/CI − 1) * 100
TSI	0.2825	28.25	(VI − 1/CI − 1) * 100
SSI	1.5414		CL/VL

where *CL* Channel length, *VL* Valley length, *VI* Valley index, *CI* Channel index, *SSI* Standard sinuosity index, *TSI* Topographic sinuosity index and *HSI* Hydraulic sinuosity index

3.2 Sinuosity and Link Channel

According to Muller (1968) there are two main factors affecting the river's sinuous hydraulogy and topography. To know the dominant factor on the formation of sinuous channel, it is essential to compute HSI and TSI (Table 2).

4 Discussion

The geomorphic set up of Krishna River and its tributaries is a clear evidence for the development of sinuosity and deviation of flow from the original course. In the field visits, it was observed that, the river was flowing in a more sinuous way in the flood plain region. Based on the slope classification given by Aswathy et al. (2008), the study area is divided into five slope classes: very gentle (0–5°), gentle (5–15°), moderate (15–25°), high (25–35°), and very high (>35°). Based on the slope analysis it is observed that, in the gentle slope areas sinuosity prevails and the channels erode the banks to shift and form a sinuous path.

5 Conclusion

It is observed that, in flood plain area of Krishna River, the sinuosity prevails which is confirmed through HSI and TSI calculated for three different channels. In case of Krishna, Panchganga and Dudhganaga the HSI is far greater than TSI. Muller (1968) also suggested that, when HSI is greater than TSI, the channel is mainly influenced by hydraulic sinuosity. Based on the sinuosity indices, it is concluded that the cutoff channels formed along the flows of Krishna, Panchganga and Dudhganaga. The SSI value is 1.54 and HSI is 71.74 which also suggests that the channel gradient has decreased relatively. This also supports the meandering channel and formation of cut-off channels at the neck of meanders.

References

Aswathy, M.V., Vijith, H., Satheesh, R.: Factor influencing the sinuosity of Pannagon River, Kerala, India: an assessment using remote sensing and GIS. Environ. Monitor Assess **138**, 173–180 (2008)

Brice, J.: Meandering pattern of the White River in Indiana—an analysis. In: Morisawa, M.O. (ed.) Fluvial Geomorphology, pp. 178–200. State University of New York Press, Binghamton, NY (1984)

Ebisemiju, F.F.: The sinuosity of alluvial river channels in the seasonally wet tropical environment: Case study of River Elmi, South Western Nigeria. CATENA **24**, 13–25 (1994)

Government of India: A report on Krishna Basin (version 2) Government of India, Ministry of water resources. www.india-wris.nrsc.gov.in (2014)

Leopold, L.B., Wolman, M.G.: River channel patterns: braided, meandering and straight. U.S. Geol. Surv. Prof. Pap. **282**, 39–85 (1957)

Muller, J.E.: An introduction to the hydraulic and topographic sinuosity indexes. 371–385 (1968). (Arizona State University temp)

Schumm, S.A., Khan, H.R.: Experimental study of channel patterns. Geol. Soc. Am. Bull. **83**, 1755–1770 (1972)

Part III

Environmental Impacts and Restoration Ecology of Natural and Agricultural Habitats

An Assessment of Environmental Impact on Agriculture in the Indus Delta Pakistan

Gohar Ali Mahar and Nayyer Alam Zaigham

Abstract

The agriculture sector in Pakistan is still the country backbone. Indus Delta was historically recognized as fertile land but the present study revealed that the severe kind of water-flow and sediment discharge from River Indus to the Deltaic area has changed the situation and deteriorated the greatest part of the agricultural land in the delta. Earlier, during the 1980s, the water flow condition of the Indus River used to be good. The finding related to the land degradation and its impact on the agricultural sector has been compiled and analyzed. (i) The results show that a large area has been submerged by the sea invasion, causing about 26% of the land of six coastal administrative units (Talukas) unusable (ii) The data collected for the period of 1998–99 to 2007–08 show extreme losses in cultivable area and more than 80% of the land was abandoned. (iii) The data collected for the same period related to the yield of rice, sugarcane and sesame also show a declining trend. (iv) Based on a survey conducted in the region, the annual earning of a farmer's family was estimated to be only $ 2 US per day. If enough water were supplied to the delta for agricultural and coastal areas or sea then the situation will be improved.

Keywords

Indus delta • Cultivated land • Indus River • Degradation • Sea intrusion

G. A. Mahar (✉)
Federal Urdu University of Arts, Science and Technology, Karachi, Pakistan
e-mail: goharmahar@gmail.com

N. A. Zaigham
GeoEnvo TechServices (GETS), Karachi, Pakistan

1 Introduction

The Indus delta is located in the south of Pakistan (Fig. 1). The farming system developed in the Indus river delta is centuries old (Mahar 2010). This region was historically recognized as fertile land with a large land area covered by a forest and the people completely reliant on agriculture. Red rice was the special commodity of the delta areas (Brohi 2003). This region was an attractive place for investors due to this fertile land, crop species and other favorable conditions (Ahmed 2004; Memon 2005). In the last four decades, fresh water scarcity and land degradation by seawater intrusion resulted in environmental deterioration and people have started to migrate from this region due to appalling environmental conditions.

Agriculture yields were high before the construction of large reservoirs and water diversions (Khan et al. 2002). Seasonal floods in the deltas provided water resources to introduce a variety of high-value crops; while silt brought down by the river have also provided fertile blankets for cropping (Brohi 2003). The Deltaic land was not only considered as one of the most fertile lands of the plain of the river Indus but also provided a lush green patch with other ecological phenomena in the whole region. Many ecological and environmental changes took place with the reduction of fresh water flow in the delta. Agriculture was one of the major affected fields due to water scarcity (Salik 2016). A maximum flow of fresh water into the sea balances the hydrodynamic level and stops seawater intrusion inland. According to the survey carried out by the irrigation department, and board of revenue, Sindh in 2000, about 326,250 (0.32 million) acre of land in the Thatta coastal district has degraded badly as a result of seawater intrusion. The main objective of this study was to find out the impact on agriculture impact and low water-flow from the upstream area because of the imbalanced hydrodynamic conditions of Indus River and seawater.

H. Chenchouni et al. (eds.), *Exploring the Nexus of Geoecology, Geography, Geoarcheology and Geotourism: Advances and Applications for Sustainable Development in Environmental Sciences and Agroforestry Research*, Advances in Science, Technology & Innovation, https://doi.org/10.1007/978-3-030-01683-8_6

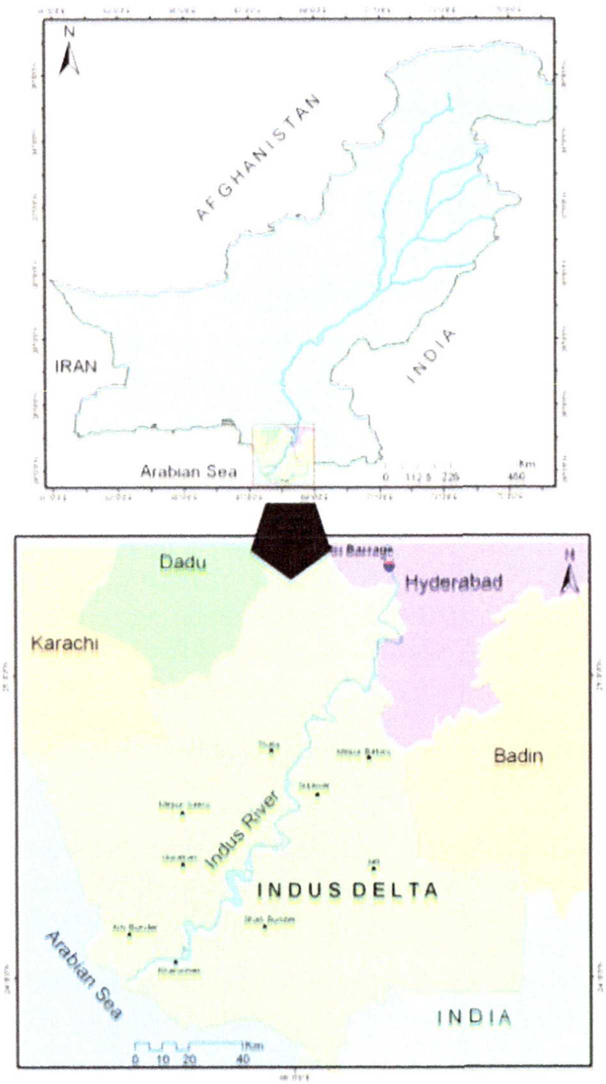

Fig. 1 Location map of the study area (Sindh, Pakistan)

2 Materials and Methods

In this study, the data collected to evaluate the environmental impact on agriculture in the Indus Delta consisted of two types primary data and secondary ones. Primary data have been collected by survey conducted in the study area and field observation. Secondary data have been collected from different sources.

The economic and social conditions of the farmers have been assessed during the field survey. The field survey was conducted in 2005–06. The radom sample selection method was used for the sampling survey and nearly 50 farmers at different locations within the study area have been selected to be interviewed (information).The pre-designed unstructured questionnaires were used to get information. The data of the collected information were compiled to evaluate (i) the cultivated average land owned by a farmer and the farmer's family structure and size and (ii) the budget of total expenditures for land cultivation and output (return) of the wheat and rice for the calculation of the net income per family.

Secondary data have been collected from different sources. Most of the data was related to the area of agricultural land, yield per hectare and production of different crops in a year. The collected data have been compiled and processed to evaluate the objective results. The data have also been plotted to evaluate the time series data and make it presentable. The collected data involved the following:

- Data about 124 Dehs (smallest administrative units), submerged by the sea intrusion.
- The recorded data of agricultural land (cultivable & uncultivable) in the period between 1998–99 and 2007–08 collected from agriculture tax (revenue) collection Department, Government of Sindh.
- Archive data of area, production and average yield of crops between 1997–98 and 2002–03 periods.

3 Environmental Impact on Agriculture

Adverse impacts on the deltaic land started after the construction of the Kotri Barrage in 1954–56, but after 1990s the deterioration rate accelerated and affected most of the cultivable lands. In the decade of 1950, progressive work for the construction of canal commands system (Dam, Barrages, Canals etc.) in the upstream area of the Indus River and was planned to improve agriculture in the country. Since then, parallel, adverse impacts on agriculture in the downstream area have started. Most of the fertile land has been transformed into saline land, due to the scarcity of water in the deltaic area. This situation has not only imbalanced the water quality at the mouth of the delta but inversely, sea intrusion has drastically destroyed most of the area. Therefore, land degradation impact on the agricultural sector has been analyzed from different view points.

3.1 Land Submerged by Sea Invasion

The six Talukas ((Boundary smaller than a district) situated near the coast of the Indus delta are Jati, Mirpur Sakro, Gorabari, Kharochan, Keti Bunder, and Shah Bunder. These Talukas comprise a total of 475 Dehs (Deh is the shortest administrative boundary in Pakistan). In 2000, the Sindh province goverment declared that out of the 475 Dehs, about 124 have been submerged (D.C.O. 2000). These results are a clear warning of the severe intensity of the sea invasion. The submergence rate of the coastal talukas is amazingly high. In

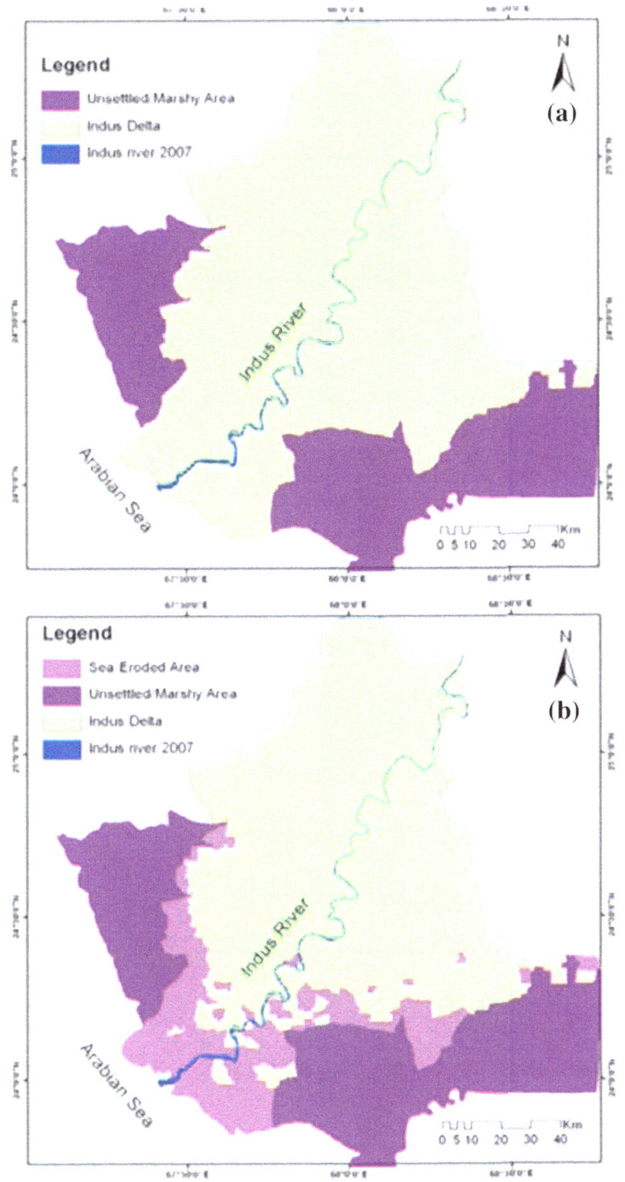

Fig. 2 Maps show administrative boundaries and migration before and after 2000. Map "**a**" represents before year-2000 marshy and the cultivated and populated area. Map "**b**" represents after year-2000, the coastal marshy, sea eroded areas from which people migrated and the cultivated area

Keti bunder about 86% of the land was submerged, 30% in Shah Bunder, 57% in Jati and about 69% in Kharochan (Fig. 2). Mirpur Sakro and Gorabari are inland talukas where the sea water impact was less severe.

3.2 Temporal Loss of Cultivated Area by Sea Intrusion

Taluka wise variations have been evaluated in the cultivated land in Thatta district. The difference of the results between the cultivated areas (during 2007–08) and the cultivable area (in the historic period) shows rigorous degradation of the Talukas, lying near the coast (Fig. 2). About 80% of the area degraded in Mirpur Sakro, 89% in Gorabari, 93% in Keti Bunder, 89% in Shah Bunder, 97% in Kharochan and 90% in Jati. The results of northern talukas of the delta do not show good condition in the cultivable areas. It was found that 60 to 70% degradation of the land has occurred in the Thatta, Mirpur Bathoro and Sajawal talukas.

The yearly loss of cultivated areas in the delta has made a discouraging scenario of the agricultural activities as shown by the graphical picture (Fig. 3). Over the last ten years, from 1998–1999 to 2007–08 our data show that the cultivated area has decreased to 35% but this decreasing trend is not constant. The total cultivable area of the district is 1,876,011 acres. In 1998–99, 25% of the total cultivable area was cultivated while in 2007–08, 16% of the total cultivable area was under cultivation. The fluctuating trend, of the loss of cultivated land of talukas, in the delta is presented in Fig. 3b. Losses are more obvious when focusing on the graphical analysis (Fig. 3c) that proves a huge difference between historically cultivable land and the land cultivated in 1998–99 and 2008.

3.3 Loss of Productivity

The scarcity of fresh water from the upstream and water intrusion from the sea have affected not only the cultivated area but also the production (yield per hectare) of crops in the delta. A decreasing tendency in production of many crops was noticed over the period 1997–98 to 2002–03 (Tables 1 and 2). It was found out from the compiled data that the production of *sugarcane*, *sesame* and *rice* has decreased at the worth seeing rate of 19, 40 and 21% respectively over the last six years. A similar impact on other crops and vegetable was also determined but at smaller rate. The production of millet, maize and cotton seemed to improve at the rate 12, 63, and 33%, respectively, during the study period.

3.4 Micro Economic Loss to Farmers

Based on the collected information, compiled and calculated, the family size of a farmer is eight people and about 3 acres of land is cultivated by the farmer. Thus, the benefits earned as 3 acre of cultivated land per 8 people. The average earnings from production of wheat and rice are 43,500/– per acre land in a year. Thus, $ 725 US (43,500 rupees Approx.) is the earning of a farmer in a year, which is about Rs. 121 or ($2) per day. This amount shows the prevalence of an extreme poverty in the deltaic area.

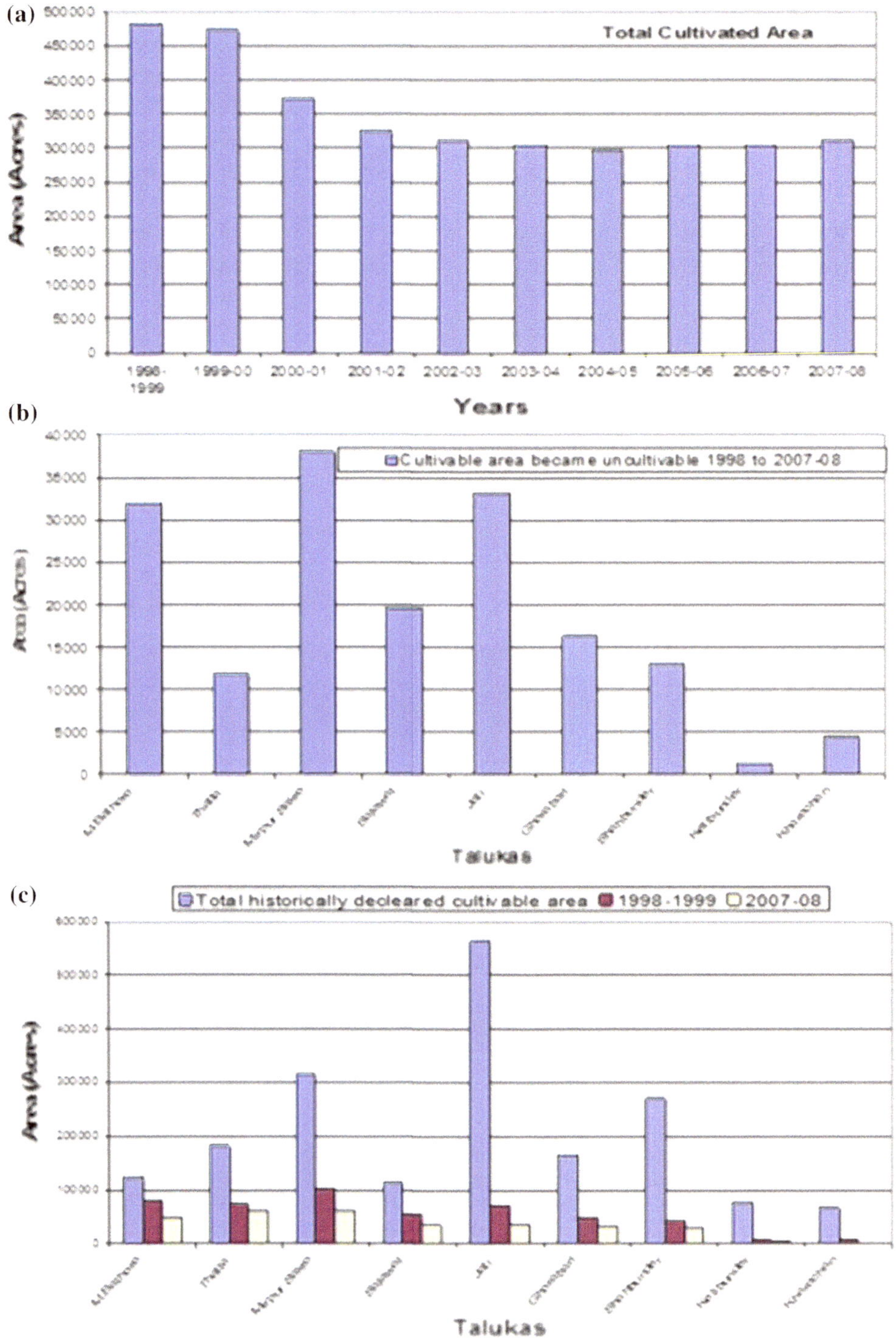

Fig. 3 Graphical representation of the total cultivable land and the cultivated land on time series data from 1998–99 to 2007–08. **a** Shows a declining trend of the cultivated land over time; **b** Shows the cultivated land decrease between 1998–99 and 2007–08 and **c** Shows the total cultivable land and cultivated land in periods of 1998–99 & 2007–08 based on administrative boundaries

Table 1 Data cultivated land (area), production and yield of crops over the periods of 1997–98 and 2002–03

Crops	Year-1997–98			Year-2002–03		
	Area (ha)	Production (mt)	Yield (mt/ha)	Area (ha)	Production (mt)	Yield (mt/ha)
Rice	65,321	130,629	1.99	56,422	89,142	1.57
Wheat	11,024	16,236	1.47	8196	11,828	1.44
Cotton	41	95 bales	2.31 bales/ha	528	1636 bales	3.09 bales/ha
Sugarcane	25,851	1,556,127	60.19	28,226	1,367,086	48.43
Jowar	178	72	0.40	427	169	0.395
Bajra	14	5	0.35	127	50	0.393
Maize	211	75	0.3 e	437	216	0.49
Sesame	43	16	0.37	62	14	0.22
Barley	8795	4540	0.44	5489	2441	0.44
Gram	408	294	0.72	23	13	0.72
Rapeseed & Mustards	1594	757	0.47	463	224	0.48

Source Bureau of Statistics Sindh, Government (1997–98 and 2002–03)

Table 2 comparative data of yield of crops over the period of 1997–98 and 2002–03 (data from Table 1), their yield in mt/ha and percentage

Crops	Yield in 1997–98 (mt/ha)	Yield in 2002–03 (mt/ha)	Difference in yield (mt/ha)	Variation (%)
Rice	1.99	1.57	−0.42	−21
Wheat	1.47	1.44	−0.03	−2
Cotton	2.31 bales/ha	3.09 bales/ha	0.78 bales/ha	33
Sugarcane	60.19	48.43	−11.76	−19
Jowar	0.40	0.395	−0.005	−1
Bajra	0.35	0.393	0.043	12
Maize	0.3	0.49	0.19	63
Sesame	0.37	0.22	−0.15	−40
Barley	0.44	0.44	0	0
Gram	0.72	0.72	0	0
Rapeseed and Mustards	0.47	0.48	0.01	2

4 Discussion

In last four decades, consequent degradation of human activities on the delta and a decline of the socioeconomic conditions of deltaic people have been the subject of discussion and research activities. Now, it is found from the recorded data that a substantial amount of water storage/divergence from the Indus River have depleted the amounts of the river water and the sediment quantity to the downstream areas. This phenomenon resulted in an environmental damage throughout the whole deltaic region. It also caused the socio-economic crises of the local community (Well and Coleman 1984; Qureshee 1999; Kahlown and majeed 2002; Kravtsova et al. 2009).

The sprawl of salinity due to the sea intrusion has severely degraded the groundwater and the land. The lack of fresh river water is the due to the huge water consumption in the Indus River upstream have reduced the cultivated land and affected the production of crops on large scale in the deltaic area. The overall environment shows continuous declining trends in the production of many crops but an improvement of production trends has also been recorded in areas associated with deltas and practically depended on the rain that fed the agricultural practices. The big loss of agricultural land with the decreasing crops production due to the fresh water scarcity of and consequently, sea intrusions have deteriorated the environmental, economic and social conditions of the farmers in the delta region. These facts were found out in field surveys. It was also found that the social values of the deltaic community are rapidly affected in response to the upstream anthropogenic activities. The huge losses of cultivated land and the productivity declining rate under the present environmental flow conditions have economically damaged the farming community of the delta region

5 Conclusion

The depletion of fresh water and sediment from upstream to the Indus delta in response to the construction of large water storage, canals network, barrages, created an environmental disparity between the fluvial and marine hydrodynamic conditions along the coastal belt, which has onset the aggressive sea wave and tidal conditions, generating devastating degradations of the land of the Indus delta. This situation has consequently affected all land use practices including cultivated land, crops production, and rangelands that have developed adverse impacts on the socioeconomic conditions of the delta people.

The cultivable land decreased and the yield and cropping trends of the cultivated crops also dropped. The farmers' income earned from an average cultivated land was found to be very low. It is concluded that the big loss of the agricultural land, the decline in the production of cash crops and staple food crops (like rice, sugarcane and wheat) had a reverse impact on the socioeconomic circumstances of the farming community of the Indus delta. Poverty has prevailed in the delta and farmers are facing extreme deprivation of basic resources for their survival and thus ensuring their existence under the minimal sustainable conditions.

References

Ahmed, F.: Freshwater resources of Indus Delta eco-region: forever Indus. In: Proceedings of the Consultative Workshop on Indus Delta 260 Eco-region (IDER), WWF Pakistan, pp. 17–36 (2004)

Brohi, S.: Livelihood resources downstream Kotri barrage and their degradation: Indus flow downstream Kotri Barrage Need orWastage, pp. 1–16. SZABIST Center for information & research, Karachi, Pakistan (2003)

D.C.O.: Land eroded by sea in the coastal Talukas of district Thatta, unpublished record of survey in 2000, 5p, District coordination office, District Thatta, Government of Sindh, Pakistan (2000)

Kahlown, M.A., Majeed, A.: Water resources situation in Pakistan. Challenges Future Strat.: COMSATS Sci. Vis. Q. **7**(3) & (4), 46–49 (2002)

Khan, T.M.A., Razzak, D.A., Qamar-uz-zaman, CH., Abdul Quadir, D. Anwarul-kabir, Sarker, M.A.,. Sea level variations and geomorpho-logical changes in the coastal belt of Pakistan. Marine Geodesy **25**,159–174 (2002)

Kravtsova, V.I., Mikhailov, V.N., Efremova, N.A.: Variations of the hydrological regime, morphological structure, and landscapes of the Indus river delta (Pakistan) under the effect of large scale water management measures. Water Resour. **36**(4), 365–379 (2009)

Mahar, G.A., Geomorphic Degradation of Indus Delta and Its Geomorphic Impact, Ph.D. Thesis, Department of Geography, University of Karachi, Karachi (2010)

Memon, A.A.: Devastation of Indus river delta. In: Proceedings of the World Water & Environmental Resources Congress, American Society of Civil Engineers, Environmental and Water Resource Institute, Anchorage, Alaska (2005)

Qureshee, M.T: Neglected coastal ecosystem of Indus delta. In: Proceedings of the National Seminar on the Mangrove Ecosystem Dynamics of the Indus Delta, collaborated by Sindh Forest and Wildlife department and World Bank, pp. 9–18 (1999)

Salik, K.M: Hashmi, M.Z.R, Waheed-ul-Zafar: Environmental flow requirements and impacts of climate change-induced river flow changes on ecology of the Indus Delta, Pakistan. Reg. Study Mar. Sci. **71**, 185–195 (2016)

Well, J.T., Coleman, J.M: Deltaic morphology and sedimentology, with special reference to the Indus river delta. In: Marine geology and oceanography of Arabian Sea and coastal Pakistan, pp. 85–100. Van Nostrand Company Scientific and Academic Editions (1984)

Conservation of Wetlands on Tunisian Islands: Kerkennah and Kuriat Islands as a Case Study

Khouloud Ben Charfi, Amjad Kallel, Thanos Giannakakis, and Imen Rais

Abstract
Mediterranean islands' wetlands have a crucial role in protecting not only the islands of the region from the effects of climate change and natural hazards, but also a wide range of endangered and endemic wildlife species, including a significant number of migratory birds. Islands' wetlands have highly dynamic changing ecosystems due to the effects of isolation and the anthropogenic perturbations that put these natural systems in a major distress and threaten their well-functioning and existence. Thus, these areas are one of the world's most threatened ones. Therefore, it is important to know, understand and monitor the changes within them to better understand the ecosystem and habitats dynamics. Through the Mediterranean Islands' Wetlands project, a methodology has been developed by the WWF Greece and disseminated to 9 Mediterranean countries in order to inventory more than 14,000 Mediterranean islands' wetlands located on 160 islands. In this project, Google earth's satellite images, national land use geo-databases with previous inventories' output were combined to identify, delineate and document possible wetlands. Together with in situ field inventory, these findings were used to better understand the evolution and changes in the systems and create a scientific database for all islands' wetlands. The aim of this study was to apply this methodology to the Tunisian islands' wetlands focusing on the significant number of wetlands that were identified in Kerkennah and Kuriat islands. This study showed the importance of building a scientific database to better understand the identified areas, and monitor their evolution in order to set up restoration and conservation strategies.

Keywords
Kerkennah archipelago • Kuriat islands
Wetlands • Salt marsh • Sebkhas • Inventory
Satellite images • Degradation

1 Introduction

The Mediterranean Basin is considered as a hotspot for islands' biodiversity; it is also rich in wetlands of both natural and artificial origins such as lagoons, ponds, salt marshes, sebkhas and salines. However, due to human activities such as the expansion of urban development, waste dumping and pollution, the wetlands of the Mediterranean islands are increasingly facing serious dangers. These activities are threatening to undermine the ecological character of the wetlands and accelerate the degradation and the ecosystem's fragmentation, which makes them extremely fragile and easily destroyed. In addition, many wetlands in the Mediterranean islands are increasingly subjected to hydrological stress, while at the same time the demand for fresh water in these areas continues to increase with the growth of both agriculture and tourism sectors.

In Tunisia, like the rest of the Mediterranean countries, environmental organizations, governments, associations or research centers are focusing on continental wetlands with important surfaces; however the islands' wetlands, with smaller surfaces, were neglected. So currently, there is a lack of information on islands' wetlands and the majority of these areas are still unidentified and undocumented, making their conservation and protection more difficult. In fact, out of 41 classified Ramsar sites in Tunisia, only 5 are located in islands (Djerba Bin El Ouedian, Djerba Guellala, Djerba Ras Rmel, Kerkennah Islands and the Kneiss Islands with their

K. Ben Charfi (✉) · I. Rais
WWF North Africa, Tunis, Tunisia
e-mail: kcharfi@wwfna.org

K. Ben Charfi · A. Kallel
National Engineering School of Sfax, Sfax, Tunisia

T. Giannakakis
WWF Greece, Athens, Greece

H. Chenchouni et al. (eds.), *Exploring the Nexus of Geoecology, Geography, Geoarcheology and Geotourism: Advances and Applications for Sustainable Development in Environmental Sciences and Agroforestry Research*, Advances in Science, Technology & Innovation, https://doi.org/10.1007/978-3-030-01683-8_7

intertidal zones) and from 256 wetlands documented during the national wetlands inventory in 1997, only 4 islands' wetlands were included (Kerkennah documented as a whole island, Kneiss, Sebkha El Quastil and sebkha of Houmt Essouk in Djerba) (Hughes 1994).

In 2004, WWF Greece launched the initiative "Conservation of the Islands Wetlands of Greece" to document and inventory the islands' wetlands of Greece, to raise awareness regarding their value and promote their conservation and protection. And in 2017, the Mediterranean Islands' Wetlands project, which is a replication of the Greek island wetland project to all the islands of the Mediterranean Basin, was set up. It includes actions for the completion of the inventories of all Mediterranean islands' wetlands, dissemination of knowledge to the public, promotion of conservation measures in local, national and Mediterranean scale and the implementation of Ramsar Resolution XII.14 "Conservation of Mediterranean Basin island wetlands".

This study, as a part of the Mediterranean Islands' Wetlands project, presented the outputs of Kerkennah and Kuriat islands' inventory.

2 Study Areas and Methods

Kuriat are two uninhabited islands, located in the Gulf of Hammamet, ca 16 km far from the coast of Monastir Bay, Tunisia. They include two main islands, the big Kuriat with a total shoreline of about 8.37 km and the small Kuriat or Conigliera with a total shoreline of 6.3 km. The distance between them is 2.3 km (Source: Satellite images, Google Earth, 2018). Characterized by terrestrial, coastal and marine ecosystems important for maintaining the biological balance of the area, and by the sheltering of some species threatened by distinction (the loggerhead turtle Caretta caretta, Pinna nobilis and the plant Caulerpa racemosa), has earned them the status of a sensitive coastal zone (following the census of biological diversity in Tunisia conducted by the Ministry of Environment and Regional Planning in 2005) and are currently part of the program of creation of Marine Protected Areas in Tunisia (APAL/SCET-TUNISIE 1999, 2000; Langar et al. 2011; CAR/ASP-PNUE-PAM 2014, 2015; Mbarek 2016).

The Kerkennah archipelago is located in the Gulf of Gabes, ca 17.9 km from Sfax, Tunisia. It includes 14

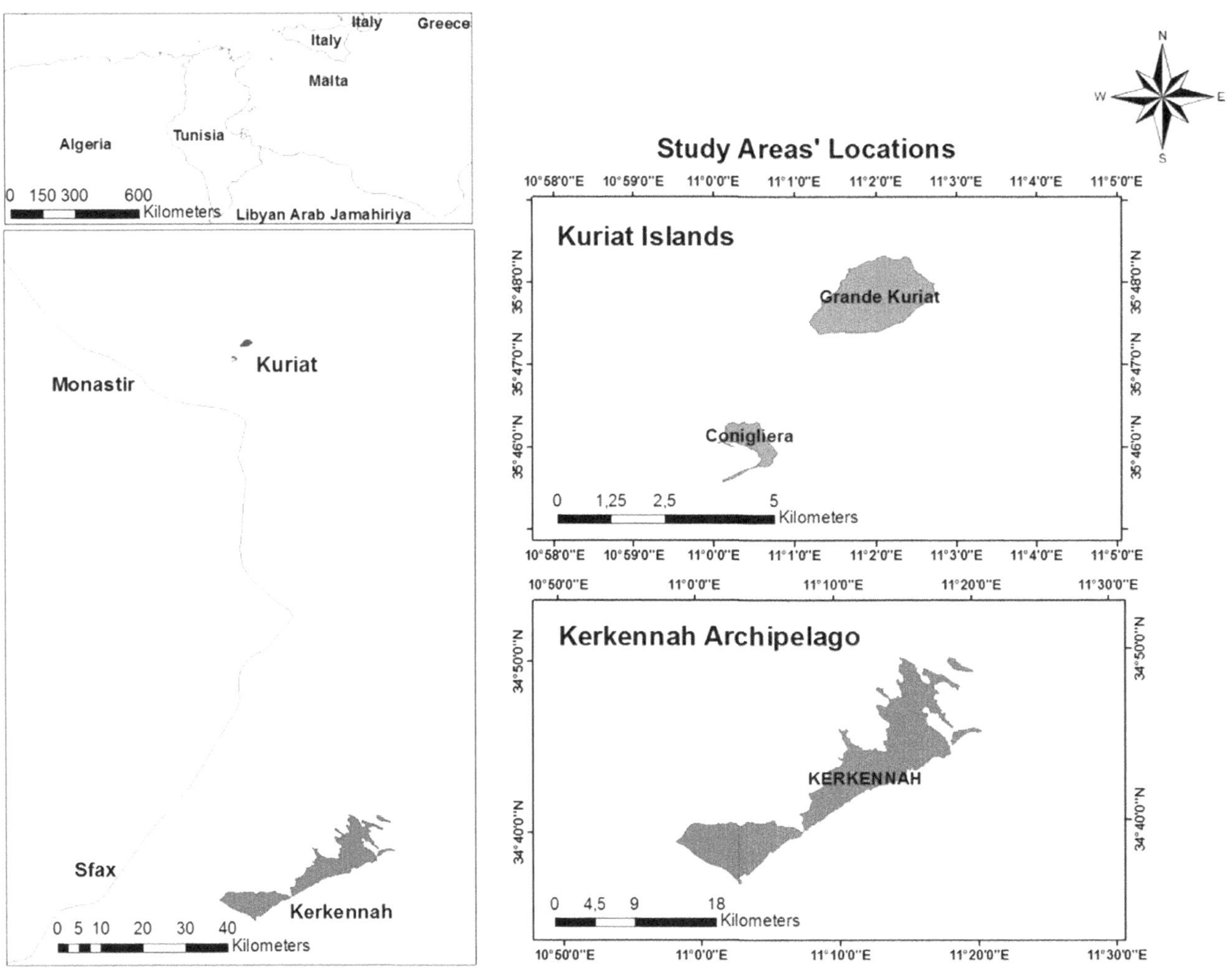

Fig. 1 Kerkennah archipelago and Kuriat islands locations

islands with a total shoreline length of 182.4 km (APAL 2015) (Fig. 1). The islands of the Kerkennah Archipelago are an important wintering site for wading seabirds and other waterbirds, as well as an important crossing site for passerines during their migrations. Plus, the satellite islets located in the northeast of the archipelago have a remarkable biological richness, with many endemic, rare or endangered animal or plant species (APAL 2001). This vulnerability, together with the significant ecological characteristics of the Kerkennah Islands, has earned them the designation of a Ramsar site and a Nature Reserve (UN-List of protected areas 2014).

The inventory strategy, developed within the framework of the conservation of Greek Wetlands' project, is based on several approaches. For the identification and the delimitation; Google Earth images, soil occupation maps and national land use geo-databases were consulted and used. Each island was thoroughly scanned and potential wetlands were marked. All wetlands, natural and artificial, were taken into account and two main criteria were set for them; non-linear Systems having a total area superior to 0.1 ha. As for the in situ approach, field visits were conducted in the spring of 2018 at a low tide period in order to find the systems not fully covered by water and be able to have a better assessment and identification of the flora species. The final mapping of the wetlands was achieved in the field and an in situ questionnaire with the locals was carried out to fully understand the history, the values and the human activities done in the area.

3 Results

A total number of 50 nonlinear wetlands larger than 0.1 ha on 14 islands were primarily identified (41 in 12 islands in Kerkennah Archipelago, 5 in big Kuriat, and 4 in Conigliera). From these wetlands, 49 were natural, only one was artificial (Salines of El Abbassia) and 29 of them were visited and inventoried.

After the field visits, 3 wetlands in Kerkennah were excluded and 9 others (5 in big Kuriat and 4 in Conigliera) were merged into 2 multi-component systems. 15.5% of the natural wetlands were highly modified or totally changed, 15.5% are partially modified, 27% are still dominated by their original habitats, and 42% are still untouched.

For the human actions, dumping of municipal waste and debris, buildings, and road construction, dikes and over-pumping, are the main noticed signs of degradation. Only one wetland, located in Kerkennah Archipelago, is under protection (designated as a Ramsar site) and no other wetland is properly managed, protected or has any kind of conservation plan. And for the last two years, 3 wetlands have been devised by the construction of dikes and are currently heavily degraded (Figs. 2, 3 and 4).

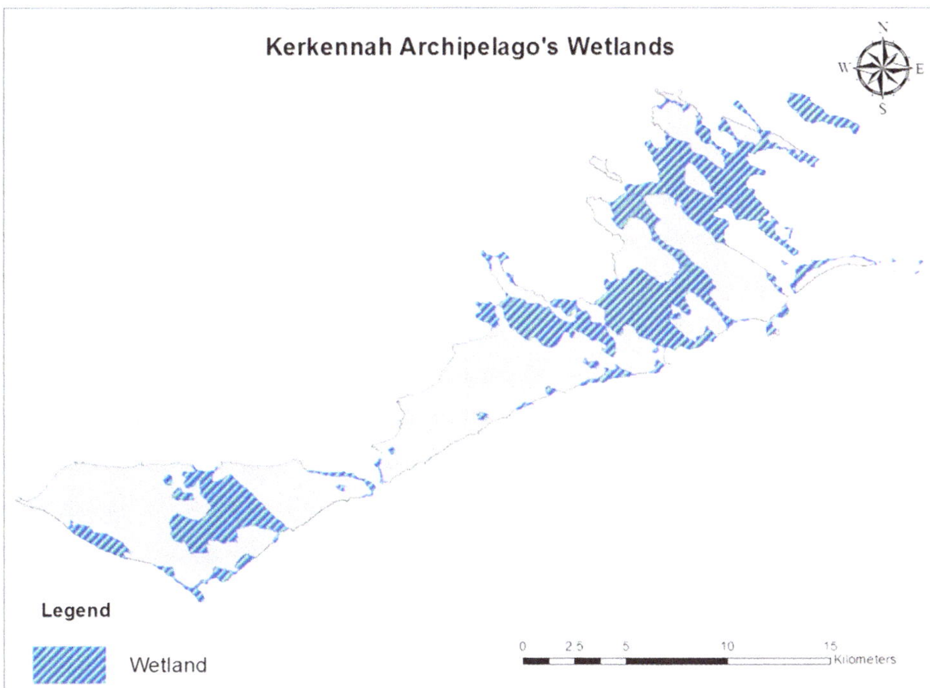

Fig. 2 Kerkennah Archipelago's wetlands

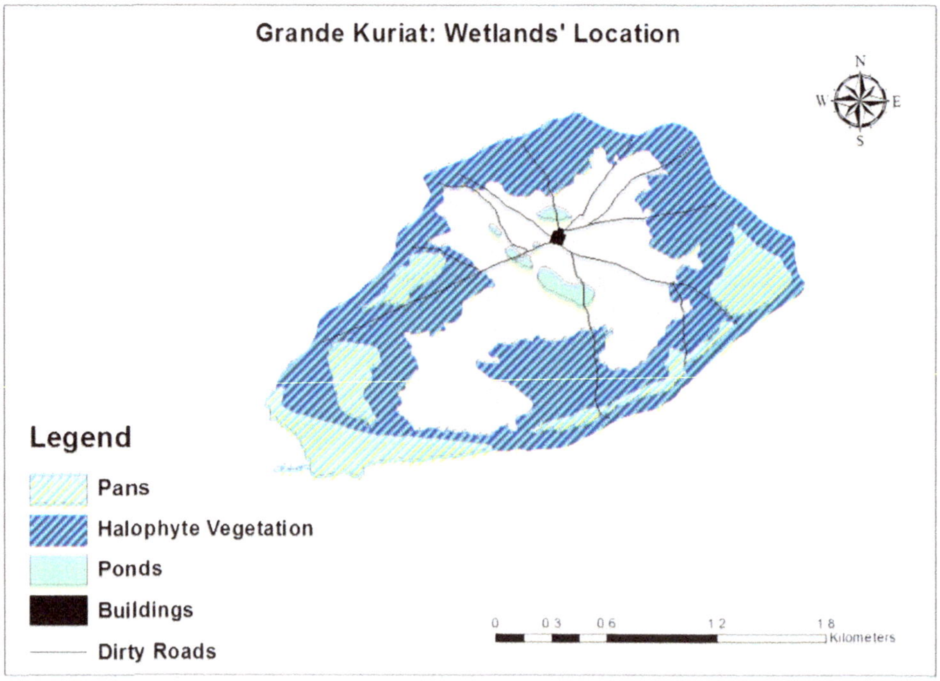

Fig. 3 Grande Kuriat's wetlands

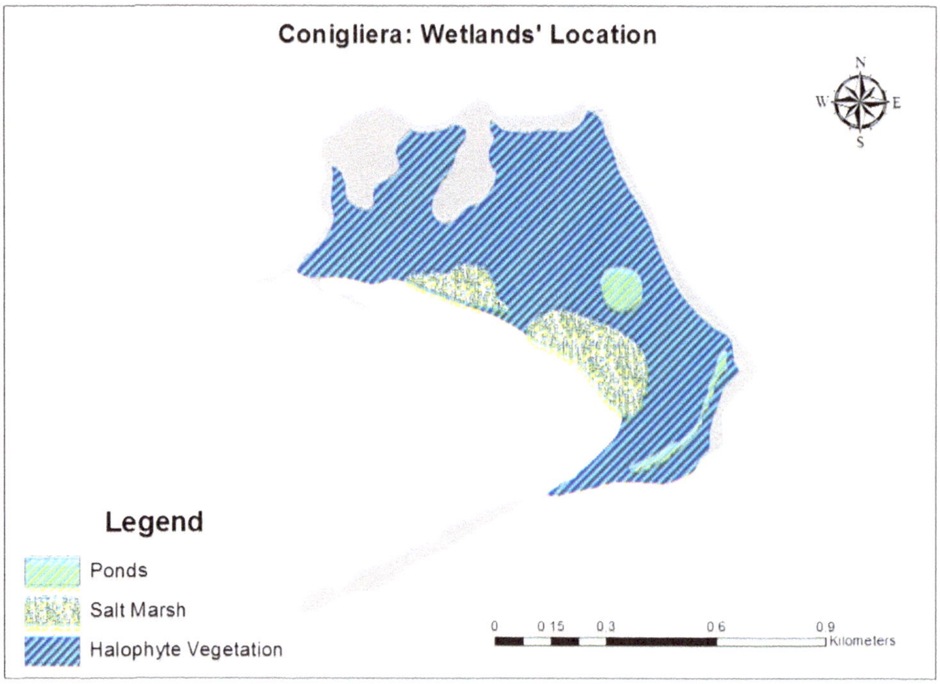

Fig. 4 Conigliera's wetlands

4 Discussion and Conclusion

The Mediterranean island wetlands are one of the most endangered ecosystems in the region. And currently, we are facing a huge gap in the documentation level of these areas which makes their conservation and protection more challenging. In this study, and through a combination of methods, we identified 50 wetlands on 14 different islands.

In Kerkennah Archipelago, two main types of wetlands were presented, coastal salt marshes and sebkhas (temporary pans). Both types are covered in halophyte vegetation and human impact on them was greatly noticed. In fact, the majority of the coastal salt marshes were used by fishermen for cabanas' building and waste dumping. As for the inlands' wetlands, the most noticed actions were chaotic household waste disposal and roads and building construction.

As for the Kuriat islands, three types were presented, coastal salt marshes, pans (salt pans and temporary pans) and ponds. Some systems are covered in halophyte vegetation and others are not. The anthropogenic impact on its wetlands, especially for big Kuriat, is limited due to their uninhabitable status.

This study shows that analyzing the status of different wetlands and establishing records of loss and degradation are fundamental for setting up management and conservation plans.

Acknowledgements This work would not have been possible without the funding of the "MAVA foundation pour la Nature", so special thanks should be given to all members of the Foundation for their constant support.Many thanks should also be given to all WWF North Africa and WWF Greece staff members and those with whom we have had the pleasure to work during this project.

References

APAL/SCET-TUNISIE: Etude de gestion des zones sensibles littorales : Iles kuriat. Technical report, Rapport de synthèse, 96 pp (1999)

APAL/SCET-TUNISIE: Gestion des zones sensibles littorales : Les îles kuriat (2000)

APAL.: *Etude de gestion de la zone sensible littorale des ilots nord-est de Kerkennah (phase 1: caractérisation du milieu naturel)* (2001)

CAR/ASP—PNUE/PAM.: Habitats marins et principales espèces des iles Kuriat (Tunisie)–Etude complémentaire: Formations naturelles d'intérêt pour la conservation. In: Par Langar, H., Bouafif, C., Charfeddine, A., El Asmi, S., Limam, A., Ouerghi, A., Sghaier, Y.R. (Ed.) CAR/ASP—Projet MedMPAnet, Tunis: 36 pages + annexes (2011)

CAR/ASP—PNUE/PAM: Elaboration d'un Plan de Gestion pour l'Aire Marine et Cotiere Protegee des iles Kuriat (Tunisie)—Phase 1: Bilan et Diagnostic. Par Thetis-Cabinet Sami Ben Haj. Ed. CAR/ASP—Projet MedMPAnet, Tunis: 72 p + annexes (2014)

CAR/ASP—PNUE/PAM: Elaboration d'un Plan de Gestion pour l'Aire Marine et Côtière Protégée des îles Kuriat (Tunisie)—Phase 2: Gestion de l'AMCP, définition des objectifs et planification des opérations. Par Thetis-Cabinet Sami Ben Haj. Ed. CAR/ASP—Projet MedMPAnet, Tunis: 56 p + annexes (2015)

Hughes, J. M., Maamouri, F., Hollis, T., Avis, C.: A preliminary inventory of tunisian wetlands (1994)

Manuels Ramsar pour l'utilisation rationnelle des zones humides, 4e édition

Mbarek, M.: Évaluation Économique des Aires Marines Protégées: Apports Méthodologiques et Applications aux îles Kuriat (Tunisie) (2016)

Vegetation Analysis of Chott Tinsilt and Sebkhet Ezzemoul (Two Ramsar Sites in Algeria) in Relation to Soil Proprieties

Adel Bezzalla, Moussa Houhamdi, and Haroun Chenchouni

Abstract

The vegetation of two Ramsar wetlands (Chott Tinsilt and Sebkhet Ezzemoul) was characterized in this research study. Plant community included 34 species belonging to 16 families in Chott Tinsilt and 43 species belonging to 17 families in Sebkhet Ezzemoul. Most of the species are halophytes with low-medium vegetation cover. The generalized linear model indicated that the soil's electrical conductivity and organic matter are the main drivers of spatiotemporal variations of plant species richness, abundance and distribution. This study proved the high plant diversity in these two wetlands, which makes a real biological heritage.

Keywords

Chott Tinsilt • Sebkhet Ezzemoul • Halophytes
Plant distribution • Diversity • Plant community

1 Introduction

Since the vegetation of a given biogeographic zone is the main ecosystem entity, expressing the main ecological conditions, the management and conservation of natural habitats, especially wetland ecosystems, implies the understanding of vegetation characteristics and its distribution in relation to the most important ecological variables (Chenchouni 2017; Neffar et al. 2018). Our current scientific understanding of plant communities characterizing North African wetlands is incomplete (Chenchouni 2012). Thus, our knowledge on trends of plant diversity of these ecosystems in this particular area remains poorly revealed. Indeed, a large scientific gap about the understanding of aspects such as ecological impact and biotic interactions and/or spatial and temporal plant distribution has been found (Boulangeat et al. 2012). In this perspective, the present study evaluated plants diversity and their spatiotemporal distribution in two Ramsar sites in northeastern Algeria (Chott Tinsilt and Sebkhet Ezzemoul). The study also investigated the spatiotemporal effect of certain edaphic factors on the variation of vegetation parameters.

2 Materials and Methods

2.1 Study Area and Vegetation Sampling Design

The two wetlands Chott Tinsilt and Sebkhet Ezzemoul (Ramsar sites) are located at province 'wilaya' of Oum-El-Bouaghi (Hauts Plateaux region of Algeria) (Fig. 1). The region has a semi-arid climate with cold winter (Aliat et al. 2016). Vegetation was sampled following the method described in Neffar et al. (2016) and Chenchouni (2017). Within each site, four stations (located at cardinal directions) were sampled every season (autumn, winter, spring, summer) of the years 2015–2016. At each station, three vegetation parameters (number of individuals, number of species, and percentage of vegetation cover) were measured in five plots of 16 m^2 (4×4 m) along a core-periphery gradient representing vegetation belts surrounding the wetland water body.

A. Bezzalla
Department of Ecology and Environment, Faculty of Natural and Life Sciences, University of Batna 2, 05078 Fesdis, Batna, Algeria
e-mail: adel.bezzalla@yahoo.fr

M. Houhamdi
Laboratoire de Biologie, Eau and Environment (LBEE), Faculty of SNV-STU, University of Guelma, Guelma, Algeria

H. Chenchouni
Department of Natural and Life Sciences, FSESNV, University of Tebessa, 12002 Tebessa, Algeria

H. Chenchouni (✉)
Laboratory of Natural Resources and Management of Sensitive Environments 'RNAMS', University of Oum-El-Bouaghi, 04000 Oum-El-Bouaghi, Algeria
e-mail: chenchouni@gmail.com

H. Chenchouni et al. (eds.), *Exploring the Nexus of Geoecology, Geography, Geoarcheology and Geotourism: Advances and Applications for Sustainable Development in Environmental Sciences and Agroforestry Research*, Advances in Science, Technology & Innovation, https://doi.org/10.1007/978-3-030-01683-8_8

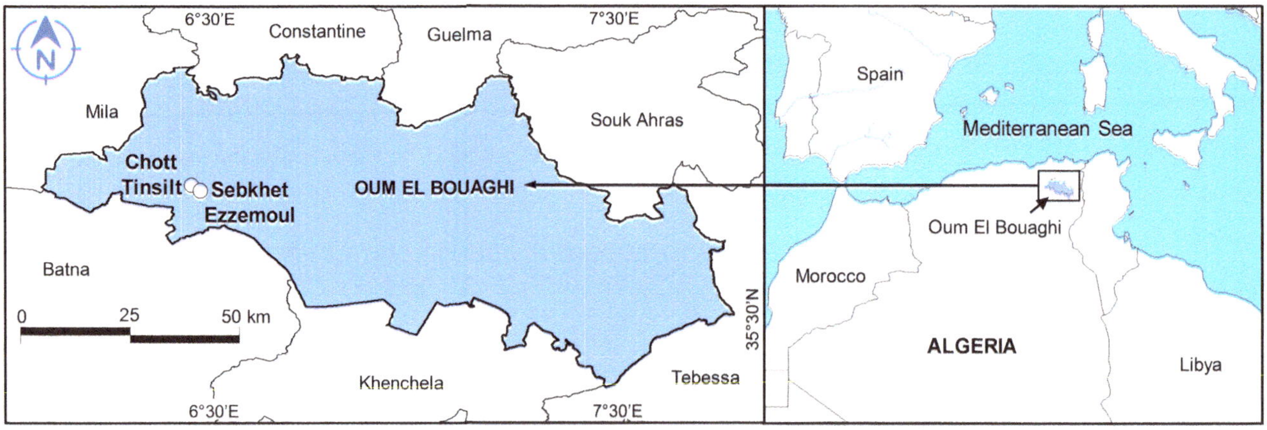

Fig. 1 Geographical location of the two study Ramsar sites, Chott Tinsilt and Sebkhet Ezzemoul, in the Wilaya 'Province' of Oum El Bouaghi, Northeastern Algeria

2.2 Data Collection and Analysis

Plant species were identified using available vegetation guides and flora identification keys of Algeria, the Sahara Desert and/or the Mediterranean region. Together with the vegetation study, soil sampling was carried out at a maximum depth of 40 cm in the five plots per station during the four seasons, i.e. a total of 80 samples were collected. For each soil sample, electrical conductivity (EC) and organic matter (OM) were determined based on the sieved soil (ø < 2 mm) using standard methods (Chenchouni 2017).

The effects of study sites, seasons, directions and vegetation belts in the interaction with soil parameters (EC and OM) on the variation of vegetation parameters (number of individuals, species richness, and vegetation cover [%]) were tested using Generalized Linear Models (GLM). Number of individuals and species richness were fitted to Poisson distribution error with log link, whereas vegetation cover was fitted to a quasi-binomial family and logit link.

3 Results and Discussion

3.1 Spatiotemporal Plant Composition and Diversity

Our floristic inventory for our two water bodies included 58 plant species, belonging to 20 different families; for the inventory at Tinsilt there were 34 species belonging to 16 families, and in Ezzemoul there were 43 species belonging to 17 families. The families Asteraceae, Brassicaceae, Poaceae, Amaranthaceae, Caryophyllaceae, and Chenopodiaceae and Fabaceae contain the highest numbers of species, whereas the majority of the other families are represented by only one or two species (https://doi.org/10.6084/m9.figshare.6291719).

The presence of four species of the family Chenopodiaceae and five species of Amaranthaceae denotes the ability of these species to resist salinity. They also have the power to resist drought after drying out. The majority of these species are C4 plants, the multiple origin of C4 photosynthesis in these two families is considered an evolutionary response to a permanent shortage in the water supply associated with high temperatures.

3.2 Vegetation Structure and Soil Effects

It is very remarkable that a succession of these species follows the affinity of each of them to endure the soil salinity. Indeed, at Tinsilt we find *Suaeda mollis* and *Suaeda fruticosa* from the first to the last row (from Belt 1 to Belt 5) after the bare mud to the periphery of the Chott, we found them in all directions of the Chott except the northern side. The western part denuded by sewage represents the largest population. Both species are highly tolerant of high salinity concentrations and also high levels of soil moisture, alternating along the transect with *Arthrocnemum indicum* and *Salicornia arabica*. Then the *Atriplex halimus* settles in belts at Belt 3, Belt 4 and Belt 5 in the South, Eastern and Western parts to give way, then, to *Silybum marianum,* a little far from the mud therefore to less salty but drier soils (Fig. 2).

At Sebkhet Ezzemoul we found *Arthrocnemum indicum* and *Atriplex halimus* from the first to the last row (from belt 1 to belt 5), in all directions of the Sebkha. *Halocnemum strobilaceum*, *Salicornia arabica*, *Hertia cheirifolia*, *Suaeda fruticosa*, *Suaeda mollis*, and *Aizoon hispanicum* have also been found throughout the year (all seasons), with smaller populations (Fig. 2). These species are highly tolerant to high salinity concentrations and also high soil moisture levels. They are alternated along the transect especially in the first

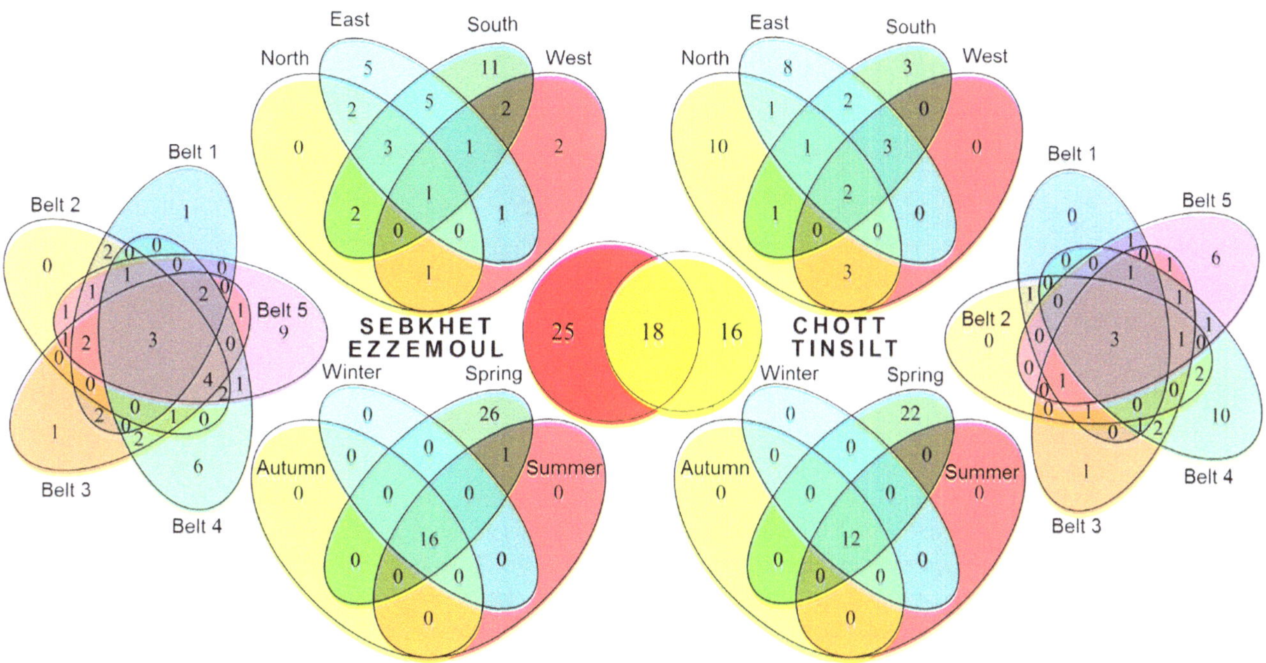

Fig. 2 Venn diagrams displaying the plant species recorded at Chott Tinsilt and Sebkhet Ezzemoul (northeastern Algeria) following site cardinal directions, seasons and core-periphery transitional vegetation belts

Table 1 Generalized linear models (GLM) testing the effects of site, season, direction and vegetation belts in interaction with soil parameters on the variation of vegetation parameters (number of individuals, species richness, and vegetation cover [%]) recorded at Chott Tinsilt and Sebkhet Ezzemoul (northeastern Algeria)

Variables	*Df*	Number of individuals			Species richness			Vegetation cover [%]		
		χ^2	*P*	Sig	χ^2	*P*	Sig	*F*	*P*	Sig
Site	1	2.35	0.125	NS	0.49	0.484	NS	1.24	0.269	NS
Season	3	27.09	<0.001	***	4.41	0.220	NS	4.27	0.008	**
Direction (Dir)	3	31.53	<0.001	***	0.65	0.885	NS	1.73	0.170	NS
Vegetation belt (Belt)	1	32.35	<0.001	***	4.59	0.032	*	7.36	0.009	**
Electrical conductivity (EC)	1	0.02	0.889	NS	0.01	0.929	NS	3.83	0.055	NS
Organic matter (OM)	1	0.90	0.344	NS	2.41	0.120	NS	2.30	0.135	NS
Site × Season	3	36.86	<0.001	***	4.13	0.248	NS	3.01	0.037	*
Site × Dir	3	20.56	<0.001	***	0.14	0.987	NS	0.46	0.713	NS
Season × Dir	9	54.08	<0.001	***	1.95	0.992	NS	1.28	0.266	NS
Site × EC	1	1.79	0.180	NS	0.19	0.667	NS	0.34	0.561	NS
Site × OM	1	0.23	0.630	NS	0.63	0.427	NS	1.30	0.258	NS
Season × EC	3	7.91	0.048	*	1.70	0.638	NS	3.92	0.013	*
Season × OM	3	34.96	<0.001	***	5.28	0.152	NS	4.28	0.008	**
Dir × EC	3	11.96	0.008	**	0.84	0.839	NS	2.37	0.079	NS
Dir × OM	3	9.01	0.029	*	1.74	0.627	NS	1.73	0.170	NS
Belt × EC	1	23.13	<0.001	***	1.43	0.231	NS	0.25	0.622	NS
Belt × OM	1	21.29	<0.001	***	1.12	0.289	NS	0.09	0.761	NS
Site × Season × Dir	9	33.02	<0.001	***	2.39	0.984	NS	0.80	0.613	NS
Site × Season × EC	3	10.57	0.014	*	0.94	0.816	NS	1.44	0.241	NS

(continued)

Table 1 (continued)

Variables	*Df*	Number of individuals			Species richness			Vegetation cover [%]		
		χ^2	*P*	Sig	χ^2	*P*	Sig	*F*	*P*	Sig
Site × Season × OM	3	33.96	<0.001	***	3.22	0.359	NS	0.96	0.416	NS
Site × Dir × EC	3	7.88	0.049	*	1.29	0.731	NS	3.60	0.018	*
Site × Dir × OM	3	2.66	0.447	NS	0.12	0.990	NS	0.75	0.527	NS
Season × Dir × EC	9	64.28	<0.001	***	1.09	0.999	NS	2.06	0.047	*
Season × Dir × OM	9	90.56	<0.001	***	2.55	0.980	NS	0.96	0.478	NS
Site × Season × Dir × EC	9	39.33	<0.001	***	4.04	0.909	NS	1.65	0.121	NS
Site × Season × Dir × OM	9	19.21	0.023	*	1.91	0.993	NS	0.51	0.863	NS

(*df*: degrees of freedom, χ^2: Chi-squared value, *F*: *F*-statistics, *P*: *P*-value; Sig.: significance codes; ***: $P < 0.001$, *: $P < 0.05$, ^{NS}: $P > 0.05$)

sampling quadrants (Belt 1, Belt 2 and Belt 3) by the species. already mentioned. A little far from the mud on less salty and drier soils we observed *Lygeum spartum* in the east, *Silybum marianum* in the south, *Thymelaea hirsuta* in the north and west, and *Limonium echoaides* in the east and west.

The list of plant species recorded in study sites showed a more or less rich and diverse flora with 58 species spread over 20 families despite the already mentioned adverse conditions including soil (texture, structure, salinity, etc.) and climatic (drought) conditions. Indeed, most of the species encountered have high tolerance towards salinity and drought. Our findings indicated that plant species are established in belts distributed in parallel along the direction of the salinity and soil moisture gradient as revealed by GLMs in combination with soil organic matter (Table 1). Among the vegetation parameters, the number of individuals (compared to species richness and vegetation cover) seems to be most influenced by the spatial and seasonal distribution of soil parameters. The edaphic and biotic factors controlling the distribution of halophytes in salty wetlands are well documented (Chenchouni 2017).

4 Conclusion

Vegetation of the two wetlands is spatially structured according to the variation of soil salinity and organic matter. The halophytic plant species were established in bands according to their threshold of tolerance vis-a-vis soil salinity and their affinity to the soil moisture level. besides, other edaphic factors, such as organic matter can intervene in the zonation of halophytes in saline environments. Species belonging to the family of Chenopodiaceae and Amaranthaceae dominate all non-flooded vegetation of both wetlands. These plant species are well adapted to saline environments.

References

Aliat, T., Kaabeche, M., Khomri, H., Nouri, L., Neffar, S., Chenchouni, H.: A pedological characterisation of some inland wetlands and Ramsar sites in Algeria. Land Degrad. Dev. **27**(3), 693–705 (2016). https://doi.org/10.1002/ldr.2467

Boulangeat, I., Gravel, D., Thuiller, W.: Accounting for dispersal and biotic interactions to disentangle the drivers of species distributions and their abundances. Ecol. Lett. **15**(6), 584–593 (2012)

Chenchouni, H.: Diversité floristique d'un lac du Bas-Sahara algérien. Acta Bot. Malacitana **37**, 33–44 (2012)

Chenchouni, H.: Edaphic factors controlling the distribution of inland halophytes in an ephemeral salt lake "Sabkha ecosystem" at North African semi-arid lands. Sci. Total Environ. **575**, 660–671 (2017). https://doi.org/10.1016/j.scitotenv.2016.09.071

Neffar, S., Menasria, T., Chenchouni, H.: Diversity and functional traits of spontaneous plant species in Algerian rangelands rehabilitated with prickly pear (*Opuntia ficus-indica* L.) plantations. Turkish J. Bot. **42**, 448–461 (2018). https://doi.org/10.3906/bot-1801-39

Neffar, S., Chenchouni, H., Si Bachir, A.: Floristic composition and analysis of spontaneous vegetation of Sabkha Djendli in north-east Algeria. Plant Biosystems **150**(3), 396–403 (2016). https://doi.org/10.1080/11263504.2013.810181

Nabkha Morphometry and Properties of Aeolian Sediments Around Native Plants in Kuwait

Ali Al-Dousari, Modi Ahmed, Noor Al-Dousari, and Safaa Al-Awadhi

Abstract

The aeolian sediments (mobile sand and dust) trapped along native plants create a geomorphological landform, known as nabkhas. About 600 core samples from aeolian accumulations settled around 15 species of main native plants in Kuwait. *Cyperus conglomeratus*, *Haloxylon salicornicum* and *Rhanterium epapposum* nabkha sediment is characterized by more alkaline (pH) values, low moisture content and electrical conductivity (EC). The later plants contain short branches spacing and are at close distances from the nearby nabkhas. The *Astragalus* sp. nabkhas mean grain sizes are considered the finest and contain considerable carbonate percentages as it grows in playa and alluvial fan deposits. *Citrulus* sp. nabkhas are characterized by the lowest pH, EC, and moisture content but the highest content of carbonate in comparison to other native plants. *Lycium* sp. has the longest distance between branches and to the equivalent nearby species. The maximum aeolian accumulation volume was noted around *Nitraria retusa*, *Lycium shawii*, *Haloxylon salicornicum* and *Calligonum polygonoides* up to 21.9, 15.5, 14.5, and 13.3 m^3, respectively, which spot them as the most effective solutions for future applications in controlling aeolian activities in the region. This study aimed to assess the sediment properties around native plants and detect the most efficient plants in controlling aeolian activities in Kuwait.

Keywords

Aeolian • Dust • Mobile sand • Nabkhas Carbonates

A. Al-Dousari (✉) · M. Ahmed · N. Al-Dousari · S. Al-Awadhi
Environment and Life Sciences Research Center (ELSRC),
Kuwait Institute for Scientific Research (KISR), P.O. Box 24885
13109 Safat, Kuwait
e-mail: Dr.Aeolian@gmail.com

1 Introduction

Aeolian processes are activated by predominant high wind speed, scanty vegetation, long drought periods, light-textured topsoil and low topographic relief (Al-Awadhi and Al-Dousari 2013). The mobile sand and dust particles movement cause sand encroachment in human settlements and poses serious economic and environmental threats (Ahmed et al. 2015). Native plants were found to have operative control measures for mobile sand in Iraq (Fadhil 2002); Saudi Arabia (Alghamdi and Al-Kahtani 2005); Iran (Pahlavanravi et al. 2012); and South Korea (Hong and Lee 2016). The native plants were found to be effective in controlling aeolian activities, soil sealing (Al-Dousari et al. 2000), wind's erosive force and act as soil organic carbon storage (Burri 2011). Native plants are proficient in catching mobile aeolian sediments forming aeolian landform called nabkha (Al-Dousari and Pye 2005). Nabkha is an aeolian accumulation around a native plant (Cooke et al. 1993; Khalaf et al. 1995; Al-Dousari et al. 2008; Ahmed et al. 2015). The nabkha formation around native plants is economically and environmentally significant as it controls mobile sand and suspended dust particles in desert and semi-desert regions (Darwish et al. 2013; Ahmed et al. 2015) and reduce sand encroachment problem. Therefore, the aim of this study was to identify the characteristics of sediments that form around dominant plants in Kuwait and detect their efficiency in controlling aeolian activities.

2 Materials and Methods

In order to test the effectiveness of each native plant species in capturing mobile sand and dust, the morphological dimensions (length, height and width) were measured through intensive field survey in 47 site locations in Kuwait (Fig. 1a, b). These native plants are *Calligonum polygonoides, Cyperus conglomeratus, Haloxylon salicornicum,*

H. Chenchouni et al. (eds.), *Exploring the Nexus of Geoecology, Geography, Geoarcheology and Geotourism: Advances and Applications for Sustainable Development in Environmental Sciences and Agroforestry Research*, Advances in Science, Technology & Innovation, https://doi.org/10.1007/978-3-030-01683-8_9

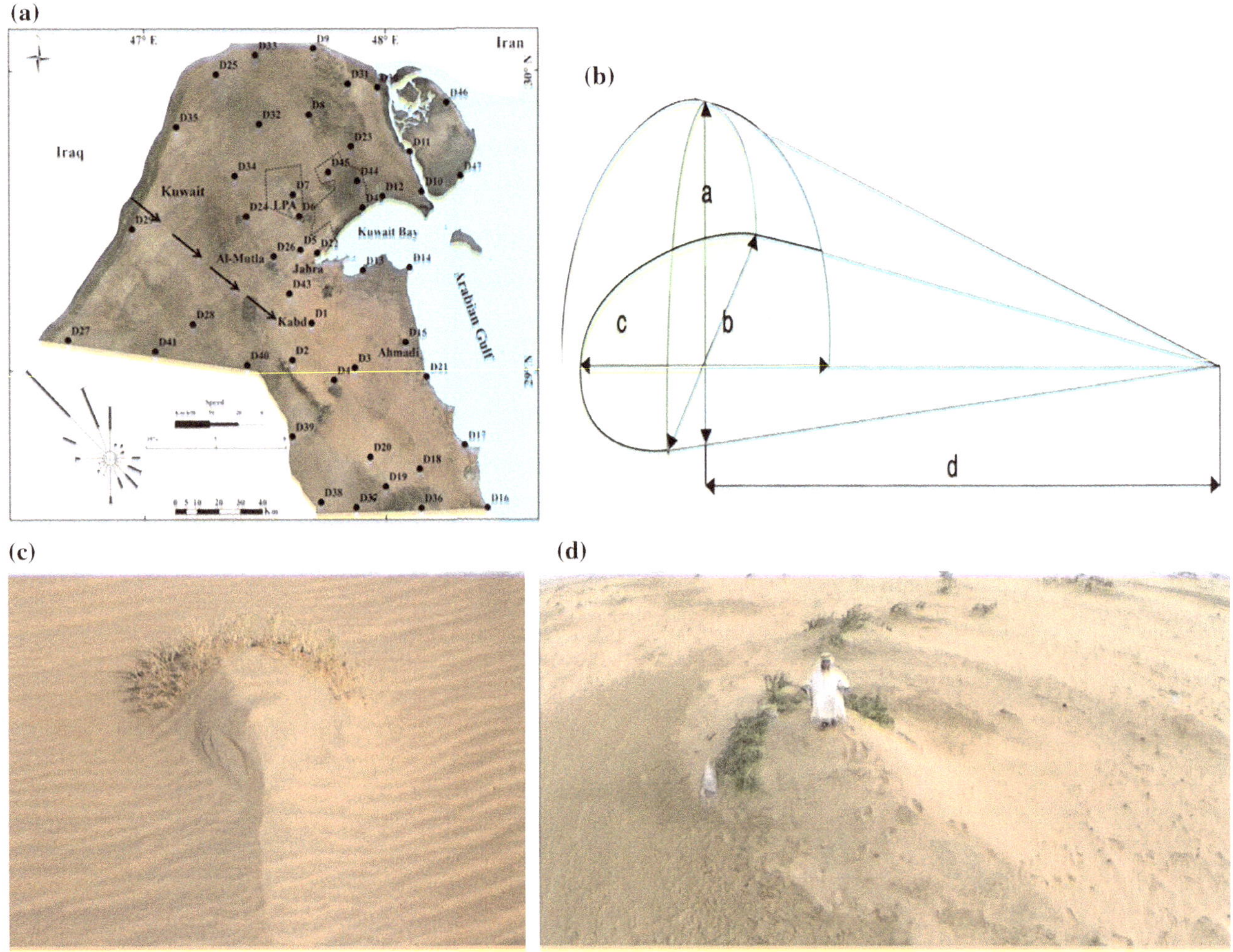

Fig. 1 Sampling location map (**a**), nabkha dimensions sketch (**b**), and field photos for nabkha of two different native plants (**c**, **d**)

Rhanteium epapposum, Astragalus spinosus, Lycium shawii, Citrullus colocynthis, Panicum turgidum, Salicornia europaea, Heliotropium bacciferum, Arnebia decumbens, Nitraria retusa, Tamarix aucheriana, Convolvulus oxyphyllus and *Halocnemum strobilaceum*. The largest nabkhas around each plant species were visually selected and measured within the 47 sites (Fig. 1c, d). In this study, the largest 20 nabkhas for each of the 15 plant species were considered for morphological dimensions and soil analysis (Alkalinity, conductivity level, moisture and particle size as well as carbonate percentages). The nabkha volume was then calculated via our formula as follows:

$$\text{The total nabkha volume} = V_E + V_P = \frac{\pi . a_1 . b_1 . c_1}{12} + \frac{\alpha . d^5 . \tan\theta}{5}$$

where a_1 is height from ground level and equal to vertical radii, b_1 is total width, and c_1 is horizontal radius. The tailing volume is calculated by integrating the base from zero to vertical radius (a) $V_P = \int A.da_1$. The area of the surface = $b_1.d$, where b_1 = d. tan θ.

3 Results and Discussion

The native plants, mainly *Nitraria*, *Lycium*, *Haloxylon,* and *Calligonum* formed the maximum volumes of nabkha deposits up to 21.9, 15.5, 14.5 and 13.3 m^3, respectively. Therefore, these plants are the most effective solutions as aeolian processes future control measures. The effectiveness of these native plants in controlling aeolian processes can be due to the morphometric parameters of these plants compared to others. This was noted by Al-Dousari et al. (2008), Ahmed et al. (2015), and El-Wahab (2018). Native plants are a workable solution to control aeolian activities in Kuwait and the surrounding areas (Al-Dousari and Al-Awadhi 2012

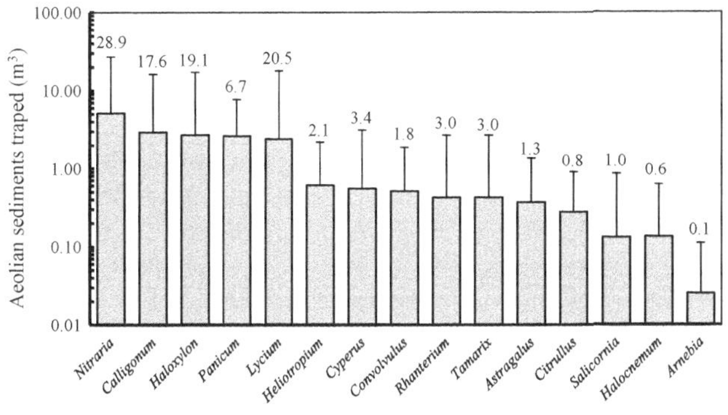

Fig. 2 The volume of aeolian accumulation around dominant native plants in Kuwait (the bars represent the maximum values)

and Al-Dousari et al. 2017). They only require 2–3 years of irrigations; then, they adapt to weather conditions (112 mm mean annual rainfall). On the other hand, aeolian sand and dust are part of the desert ecological system and they act as a healing factor for degraded lands as a constant supplier of plants seeds. Dust and mobile sand contain organic matter up to 8.96 and 4.85%, respectively from total sample weight (Al-Dousari et al. 2016). The proper use of native plants in controlling aeolian mobile sand was successful in Syria (Al-Obaid 2000), Saudi Arabia (Alghamdi and Al-Kahtani 2005), Iran (Pahlavanravi et al. 2012), Chad (Djibril 2013), Egypt (Darwish et al. 2013), Morocco (Hammouzaki 2013) and China (Hong and Lee 2016).

The *Astragalus* sp. nabkhas mean grain size are considered to be the finest and high in carbonates as it grows in playa and alluvial fan deposits. *Citrulus* sp. nabkhas are characterized by the lowest pH, EC, and moisture content but the highest content of carbonate in comparison to other native plants. *Lycium* sp. has the longest distance between branches and to the equivalent nearby species. *Cyperus conglomeratus*, *Haloxylon salicornicum* and *Rhanterium epapposum* nabkha is characterized by more alkaline (pH) values, low moisture content and electrical conductivity (EC). The latter plants develop short branches spacing and at close distances with nearby nabkhas (Fig. 2).

4 Conclusion

The average aeolian accumulation around native plants in Kuwait is equivalent to 1.92 m^3. Some native plants show the best capability in controlling aeolian sediments. The maximum volumes of aeolian accretion were noted to be around *Nitraria retusa* (21.9 m^3), *Lycium shawii* (15.5 m^3), and *Haloxylon salicornium* (14.5 m^3). It is noteworthy to consider them as the most efficient plants in controlling aeolian sediments in the region. Correspondingly, each native plant has its distinctive morphological and soil characteristics. The native plants were found to be the practical solutions in controlling aeolian accumulations and as they have a remarkable low cost. Therefore, they are recommended as potential future applications to control sand encroachment and resuspended dust particles in the region.

Acknowledgements Gratitude for supporter (KISR) Grant number: 2011/EC085C. Appreciation is extended to Abu Noor, Mamdouh and Sayed for their helping with laboratory and field measurements.

References

Ahmed, M., Al-Dousari, N., Al-Dousari, A.: The role of dominant perennial native plant species in controlling the mobile sand encroachment and fallen dust problem in Kuwait. Arab. J. Geosci. **9** (2), 134 (2015). https://doi.org/10.1007/s12517-015-2216-6

Al-Awadhi, J.M., Al-Dousari, A.M.: Morphological characteristics and development of coastal nabkhas, North-East Kuwait. Int J Earth Sci. (Geol Rusdsch) **102**, 949–958 (2013). https://doi.org/10.1007/s00531-012-0833-9

Al-Dousari, A., Doronzo, D., Ahmed, M.: Types, indications and impact evaluation of sand and dust storms trajectories in the Arabian Gulf. Sustainability **9**(9), 1526 (2017). https://doi.org/10.3390/su9091526

Al-Dousari, A., Misak, R., Shahid, S.: Soil compaction and sealing in Al Salmi area. J Land Degradation Dev **11**, 401–418 (2000). https://doi.org/10.1002/1099-145X(200009/10)11:5%3c401:AID-LDR396%3e3.0.CO;2-4

Al-Dousari, A.M., Aba, A., Al-Awadhi, S., Ahmed, M., Al-Dousari, N.: Temporal and spatial assessment of pollen, radionuclides, minerals and trace elements in posited dust within Kuwait. Arab. J. Geosci. **9**, 95 (2016). https://doi.org/10.1007/s12517-015-2182-z

Al-Dousari, A.M., Ahmed, M., Al-Senafy, M., Al-Mutairi, M.: Characteristics of nabkhas in relation to the dominant perennial plant species in Kuwait. Kuwait J. Sci. Eng. **35**(1), 129–150 (2008)

Al-Dousari, A.M., Al-Awadhi, J.: Dust fallout in northern Kuwait, major sources and characteristics. Kuwait J. Sci. Eng. **39**(2A), 171–187 (2012)

Al-Dousari, A.M., Pye, K.: Mapping and monitoring of dunes in northwestern Kuwait. Kuwait J. Sci. Eng. **32**(2), 119–134 (2005)

Alghamdi, A.A., Al-Kahtani, N.S.: Sand control measures and sand drift fences. J. Perform. Constructed Facil. **19**(4), 295–299 (2005). https://doi.org/10.1061/(ASCE)0887-3828

Al-Obaid N.: Economic and environmental assessment for direct and indirect factors causing desertification (base study in Bushra-Syria). The Arab Center for the Studies of Arid zones, Syria (2000)

Burri, K.: Plant and mycorrhizal Fungi in wind erosion control. Ph.D. Dissertation, Swiss Federal Institute of Technology in Zurich (ETH Zurich) (2011)

Cooke, R.U., Warren, A., Goudie, A.S.: *Desert Geomorphology*, pp. 357. Taylor & Francis, London (1993). ISBN 1-85728-017-2

Darwish, K., Safaa, M., Momou, A., Saleh, S.A.: Egypt: Land degradation issues with special reference to the impact of climate change. In: Heshmati, A.G., Squires, R.V. (eds.) Combating desertification in Asia, Africa and the Middle East: Proven practices, pp. 113–136. Springer, Netherlands, Dordrech (2013)

Djibril, H.: Desert environments of republic od chad. In: Heshmati, A. G., Squires, R.V. (eds.) Combating desertification in Asia, Africa and the Middle East: Proven practices, pp. 169–189. Springer, Netherlands, Dordrech (2013)

El-Wahab, R.A., Al-Rashed, A.R., Al-Dousari, A.: Influences of physiographic factors, vegetation patterns and human impacts on aeolian landforms in arid environment. Arid Ecosyst. **8**(2), 97–110 (2018)

Fadhil, A.M.: Sand dunes fixation in Baiji, Iraq. J. China Univ Geosciences **13**(1), 67–72 (2002)

Hammouzaki, Y.: Desertification and its control in Morocco. In: Heshmati, A.G., Squires, R.V. (eds.) Combating Desertification in Asia, Africa and the Middle East: Proven practices, pp. 91–110. Springer, Netherlands, Dordrech (2013)

Hong, S.H., Lee, E.Y.: Restoration of eroded coastal sand dunes using plant and soil-conditioner mixture. Int. Biorem. Biorem. **113**, 161–168 (2016). https://doi.org/10.1016/j.ibiod.2016.04.021

Khalaf, F.I., Misak, R., Al-Dousari, A.M.: Sedimentological and morphological characteristics of some nabkha deposits in the northern coastal plain of Kuwait Arabia. J. Arid Env. **29**(3), 267–292 (1995). https://doi.org/10.1016/S0140-1963(05)80107-7

Pahlavanravi, A., Miri, A., Ekhtesasi, M.R.: The impact of different kinds of dust storms in hot and dry climate, a case study in Sistan region. Desert **17**(1), 15–25 (2012). http://jdesert.ut.ac.ir

The Ecological Role of *Cornulalca aucheri* (Amaranthaceae) in the Stabilization of Degraded Sandy Soils in Kuwait: The Case Study of Liyah Area

Modi M. Ahmed, Noor Al-Dousari, and Ali M. Al-Dousari

Abstract

Cornulaca aucheri is an annual and biannual herb considered as a disturbance indicator currently visible and widely distributed in disturbed lands in Liyah area in Kuwait. Such area suffers from severe land degradation due to multiple interacting factors such as, overgrazing, camping and recreation, gravel and sand quarrying, military activities and natural processes. A restoration program was applied after the quarry surface irregularities were refilled and levelled. During the past 10 years of rehabilitation, noticeable healthy cover of *Cornulaca* sp. has been seen especially around an artificial lake and playas. The existence of such species in high density means that the restoration program was a success and was transitted from bare ground to *Cornulaca* and annual forb state. This is a lower state of Range State Transition Succession model, but it is better than bare soil. *Cornulaca* sp. is a native desert plant that grows in arid conditions. Despite its spiny leaves, it provides good food for camels and increases the supply of milk produced by lactating females. It is about 80 cm tall and has a stem that is branched from the base with new faster greenery growth in the summer. It shows good environmental potential to be managed as natural types used for restoration of degraded lands in desert areas.

Keywords

Land degradation • Range • State transition
State rehabilitation

M. M. Ahmed (✉) · N. Al-Dousari · A. M. Al-Dousari
Crisis Decision Support Program/ELSRC, Kuwait Institute for Scientific Research, P.O. Box: 24885, 13109 Safat, Kuwait
e-mail: mmahmed@kisr.edu.kw

1 Introduction

Omar (2000) proposed the Range Succession and State/Transition Model (RSST) (Fig. 1). At the upper level of the RSST model, the perennial species dominate and, the range conditions improve. In contrast, annual species overlook and range condition decline at the lower levels of the model. The succession state at Liyah area starts from the lower level (bare land) but after a ten-year protection and restoration, the level shifts to dominant of *Cornulaca* and annual forbs state. The section below high lights the annual plant *Cornulaca aucheri*.

The area of Liyah is about 179 km^2 (1.12% of total area of Kuwait). It is almost flat terrain interrupted by a group of hills including Liyah Ridge (approximately 100 m above sea level), dry valleys and depressions (playas). The abundance of aeolian and gravel deposits led to the gravel extraction activities in the early sixties of the last century (Alhajerif et al. 2008). The soil in the northern part of Liyah area is deflated showing coarse sand or granules and rich with calcium carbonate. The study area suffers from soil crusting, compaction, soil loss, missing silt from the soil, low density of the vegetation cover and loss of biodiversity, due to the extensive gravel extraction activities. The condition of the area was further exacerbated by the rainfall scarcity, intense wind and water erosion.

2 Methods

To achieve our research purposes the following tasks were carried out: Surveying the dominated native plants to identify the range condition. Identifying the morphological parameters (length and width of plant and height of sand body) for 20 Nabkhas of *Cornulaca* plants. Measuring the depth and direction of the roots for five *Cornulaca* plants. Measuring the pH and electrical conductivity for 20 *Cornulaca* nabkhas parameters.

H. Chenchouni et al. (eds.), *Exploring the Nexus of Geoecology, Geography, Geoarcheology and Geotourism: Advances and Applications for Sustainable Development in Environmental Sciences and Agroforestry Research*, Advances in Science, Technology & Innovation, https://doi.org/10.1007/978-3-030-01683-8_10

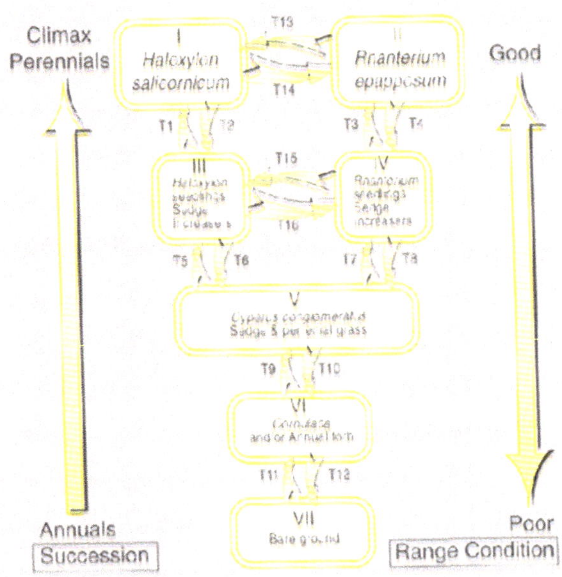

Fig. 1 The succession model for *Haloxylon* and *Rhanterium* community types (Omar 2000)

3 Results and Discussion

In dry seasons the territory of the study area becomes brown, and the majority of the plants appear to be dead. Only *Cornulaca* plants are green and healthy. *Cornulaca* community is one of the most abundant shrubs in rehabilitated Liyah area. They are considered as indicators of land degradation, and poor ingenuity type of annual shrub colonizes the smooth sand sheet about >10 cm thick. The preferred sites for this plant are wadis, depressions, and playas. The plants have spread to areas that were previously dominated by perennial plant such as, *Rhanterium epapposum, Haloxyon salicornicum*, and *Cyperus conglomerates*. This plant grows, blooms and seed sets in the favorable conditions in the summer season. It has stems that branch from the base with new growth in summer, smooth spiny leaves that become stiffer as the plant matures.

It is annual or short-lived perennial, 94.2 cm in average length at the artificial lake and 141.7 in average height at the playa (Table 1), and succulent branched from base. Leaves are linear-filiform to triangular-subulate, 4–10 mm long, greyish-green, recurved, white spine and densely woolly in axils. Flowers 1–8, in axils of leaf-like bracts, surrounded by white hair. It has the ability to form medium sized nabkha. It can be used for sand fixation and for control measures because it can trap sand. The average heights of sand body at the artificial lake and at playa are 67.1 and 82.9 cm, respectively. The morphological measures of *Cornulaca* sp. at playa are larger than around artificial lake. The soils in playa are loosely consolidated than the soil around the artificial lake allowing the deep penetration of roots.

Such plants grow in alkaline soil having a pH value between 7.2 and 7.9 and low electrical conductivity from 2.4 to 3 $\mu S\ cm^{-1}$. Thus there is an adaptation to survive in an arid environment through developing many physical and behavioral mechanisms similar to desert animals. For example, their stem is vertically-erect, coated with waxy water proof cuticles, green that maximize light interception during the early and late hours of the day but avoids the midday sun, because excessive heat may damage the plant tissues and have special fleshy tissues to conserve and store water (Hewitt 1993). The leaves are linear to reduce transpiration and recurved to protect from the wind action and to withstand heat and bright. Their leaves have special modification, they ended with a silver spine and stomata chamber covered with hair to reflect the light and the heat. The root

Table 1 The morphological dimensions of *Cornulaca* sp. around the artificial lake and playa in Liyah, Kuwait

No.	Artificial lake			Playa		
	Length (cm)	Width (cm)	Height of the sand body (cm)	Length (cm)	Width (cm)	Height of the sand body (cm)
1	143	147	68	130	120	107
2	130	149	84	133	133	80
3	79	80	71	130	185	103
4	102	100	67	122	128	82
5	111	135	58	154	155	73
6	92	99	81	147	142	71
7	85	90	72	157	155	70
8	60	88	65	122	105	67
9	70	72	47	167	125	94
10	70	90	58	155	142	82

Table 2 Depth of roots (cm) of *Cornulaca* sp. playa, Liyah

No	Length of root (cm)	Direction
1	55	Vertical
2	50	Vertical
3	100	Horizontal
4	70	Vertical
5	90	Vertical

system sustains growth in dry periods by extending either horizontally or vertically according to water availability in the soil, thus offsetting the effects of drought. This root system can also penetrate deeply into the soil about one meter to reach moisture level (Table 2). These shrubs can survive in the dry ecosystem due to their ability to harvest rapidly and efficiently from limited water resources, such as mist and dew. Habitats of these water preserving plants are often in areas with high temperatures and low rainfall. In addition many native plants prefer to grow inside or near *Cornulaca* shrub. It supplies food and shade for other plants.

4 Conclusion

Desert shrub plant can be used for rehabilitation programs in degraded soil for the following socioeconomic, environmental and ecological criteria (SER news 2016):

- the capacity to grow in nutrient-poor soil,
- the ability to grow on the disturbed area, undulating sandy plains, sand sheet, gravelly plains, rocky debris, top and slopes of sand accumulation and off roads,
- the modification of root system (horizontal or vertical) that can quickly reach the residual soil moisture,
- they are equipped to survive in an ecosystem that contains scarce water sources,
- the rapid growth in summer season and the ability to regenerate easily,
- the potential to control shifting sands through the development of small nabkha,
- the ability to resist strong wind, hot temperature and lightness during the day,
- the capacity to provide forage for wildlife and livestock especially for camels,
- the capability to assist in protection of soil against movement by wind or water and,
- they have a high artistic value in bringing diversity and color to the arid landscape.

References

Alhajraf, S., Al-Enezi A., Al-Dousari, A.M., Brown G., Delima, E.: Environmental rehabilitation study of the quarries in the State of Kuwait, Al-Liyah. No. KISR 9475. Kuwait Institute for Scientific Research, Kuwait (2008)

Hewitt, T.: The complete book of Cacti & Succulent. Covent Garden Books, London. (1993) ISBN 978-1-85605-402-7

http://ser.org/resources/sernews. Restor. Ecol. J. 30(3) (2016). Last accessed on 25 June 2016

Omar, A.S.: Vegetation of Kuwait, pp. 16. Kuwait Institute for Scientific Research (KISR), Kuwait (2000)

Orchard Understory Species Diversity in Relation to Orchard Age and Soil Factors (in Ibadan, Southwestern Nigeria)

Olutoyin Adeola Fashae, Adeyemi Oludapo Olusola, Arioluwa Aribisala, and Aishah Oyeniyi

Abstract

The relationship between species richness and environmental variables enhances valid conclusions on the effect of biodiversity on ecosystem processes. Extant literatures exist on the relationship between species richness, a measure of biodiversity and ecosystem functionality. Therefore, this study examined understory plant species diversity of Citrus orchards (*Citrus sinensis*) between the young (twenty years) and the old (forty years) in relation to soil parameters. Direct method of gradient analysis was used whereby sample plots were randomly located and marked at the study site and understory plant species per unit area recorded and categorized. A set of 30 quadrats and 39 quadrats were taken in the young and old orchards respectively. The collected data in the undergrowth were subjected to an ecological analysis. The data were analyzed using Simpson's index, student T-test and canonical correspondence analysis (CCA). The result of the diversity index showed that the older orchards had greater species dominance than the younger ones. The t-test shows a significant difference between orchards ($t = 0.03$, $P = 2.27$). The placement of environmental variables in CCA showed that the first gradient (that is, the main explainable variation in floristic composition) was positively correlated with pH, total nitrogen and negatively with clay content.

Keywords

Plant diversity • Undergrowth • Citrus orchard Species richness

1 Introduction

Ecological communities function and are sustained mostly based on the health of species diversity. A number of factors affect small-scale species diversity, including environmental factors such as disturbance, environmental heterogeneity and resource availability; geographic factors such as the regional species pool, dispersal distance and ease of dispersal, as well as biological factors such as predation, competition and facilitation. The significance of these factors varies depending on the scale of observation. Also, small-scale diversity can influence features of ecosystem function including productivity and stability. The pattern of species diversity along the environmental gradient has been a subject of inquiry in biogeography (Fashae et al. 2018). Quite a huge number of experiments has been performed to check out the relationship between species richness (a measure of biodiversity) and ecosystem function. Basically, understanding the relationship between species richness and environmental variables helps us to draw valid conclusions on the effect of biodiversity on ecosystem processes especially across orchards of different ages (Naffar et al. 2013; Meddad-Hamza et al. 2017). Therefore, this study examined plant species diversity of undergrowth under citrus orchards (*Citrus sinensis*) of ages between 20 (young) and 40 (old) years in relation to soil nutrient.

2 Study Area

The study site is located within the National Horticultural Research Institute (NIHORT), Ibadan, Oyo State, Nigeria. The Institute is located at Latitude 7° 23′ and 7° 25′N and Longitude 3° 50′ and 3° 52′E. It is situated on a 350 ha of land. NIHORT is located within Ibadan and falls within the humid tropical zone which covers about 21% of the total landscape of Nigeria (Iloeje 1981). There are two main climatic seasons: tropical wet and dry climate with a lengthy

O. A. Fashae (✉) · A. O. Olusola · A. Aribisala · A. Oyeniyi
Department of Geography, University of Ibadan, Ibadan, Nigeria
e-mail: oa.fashae@mail.ui.edu.ng; toyinafashae@yahoo.com

H. Chenchouni et al. (eds.), *Exploring the Nexus of Geoecology, Geography, Geoarcheology and Geotourism: Advances and Applications for Sustainable Development in Environmental Sciences and Agroforestry Research*, Advances in Science, Technology & Innovation, https://doi.org/10.1007/978-3-030-01683-8_11

wet season and relatively constant temperatures throughout the course of the year. The sites are well drained by two rivers. The dominant soil types are the ferruginous group of tropical soils. Agricultural land use type in the area is mainly for the production of horticultural crops such as vegetables, fruits and ornamentals. there are orchards of *Magnifera indica* (Mango), *Citrus* spp. (Orange, Lemon, Lime, Grapes and Mandarins), *Treculia africana* (Bread Fruit), *Psidium guajava* (Guava), *Chrysophylum albidum* (African star apple) *and Irvingia gabonensis* (Bush Mango). In the dry season, vegetable production is extensively practiced in the valley bottom. The young citrus orchard area extent is about 4659.4 m^2 while the old citrus orchard area extent is about 7555.5 m^2. The citrus is planted along 9 rows, 7 m apart both horizontally and vertically.

3 Methods

The gradient analysis using sample plots randomly located and marked at the study site was performed. Understory plant species per unit quadrat (50 cm × 50 cm) were recorded and categorized. The geographic positions of the quadrat points were taken and mapped. A total of thirty and thirty-nine quadrats were taken in the young and old orchards respectively. Data collected in the undergrowth were subjected to an ecological analysis. After collecting the vegetation data, a volumetric soil sample at 0–30 cm depth was collected randomly with a soil auger at the center of each sampled quadrat. Soil physicochemical parameters of pH, organic carbon, organic matter, total nitrogen, copper, zinc, magnesium, phosphorus and particle size distribution were analysed. The data were analysed using Simpson's index, student T-test and canonical correspondence analysis (CCA). For interpretation (CCA), each site was labelled with reference to its age. The points labelled Y and O are the points within the understory of the young and older orchard, respectively. The length of the environmental arrows indicates how important that variable is and how well the values of the variable are displayed in the biplot of the environmental variables and sites. This follows the fact that the length is equivalent to the maximum rate of change of the variable; across the diagram, variables with short arrows have little variation.

4 Results

The understory plant species found under the younger citrus orchard were grouped into 25 families with about 59 species. Out of these, the families with the highest dominance in terms of number of species were *Poaceae* (29%), *Asteraceae* (27%), *Fabaceae* (18%), *Amaranthaceae* (13%) and *Acanthaceae* (13%). As expected these families dominated the younger orchards because they were widespread under cultivated lands and field crops. The older citrus orchards had 61 understory species. The dominant species within the older orchard in terms of dominance are *Setaria barbata, Euphorbia graminea, Oplismenus burmannii, Althernanthera brasiliana and Asystasia gangetica*. The families with the highest dominance in terms of number of species were *Poaceae* (34%), *Acanthaceae* (22%), *Amaranthaceae* (13%), *Euphorbiaceae* (10%), *Asteraceae* (8%), *Vitaceae* (7%), *Fabaceae* (6%), and *Acanthaceae* (13%). The result of the Simpson diversity index showed that the older orchard had greater species richness (0.91) than younger orchard (0.88). The *t*-test shows a significant difference of 0.03 (p = 2.27). Implying that there is a significant difference between the means of the two ages of the orchards.

The placement of environmental variables shows that the first gradient (that is, the main explainable variation in floristic composition) is positively correlated with pH, total nitrogen and negatively with clay content percentage. Likewise, some species composition had a negative correlation with Fe and Mn and a positive correlation with Zn, Na and other elements. The species points and the environmental arrows together represent weighted averages, typically giving a summary of the species along each of the environmental variables. By projecting the species points on the arrows for pH (Fig. 1), we can infer that *Althernanthera brasiliana* and *Centrosema pubescens*, of all the displayed species, have the highest weighted averages for pH and for this reason occur at high pH values and *Talinum triangulare* at lower than average values. On the other hand, species close to *Tridax procumbens* have low weighted average for Mg and Avail. P.

5 Discussion

The species count between the young and old orchards is quite close but differences exist between species across the orchards. The Simpson index suggests significant difference between their diversities. It was observed that there were endemic species in each of the orchards. The result of the t-test implies that age significantly influenced understory diversity. This suggests that age affects the species diversity of understories to a large extent. Newer species are colonizing the younger orchards as a result of continued cultivation and management of the younger understory. As regards the canonical correspondence (Fig. 1), it is possible to infer relative abundance from the diagram but how exactly this should be done is still a topic for discussion in literature (Ter Braak et al. 1994). We therefore observed that sites that contain a particular species are spread around the point of that species. In Fig. 1, *Ageratum conyzoides* is in the middle

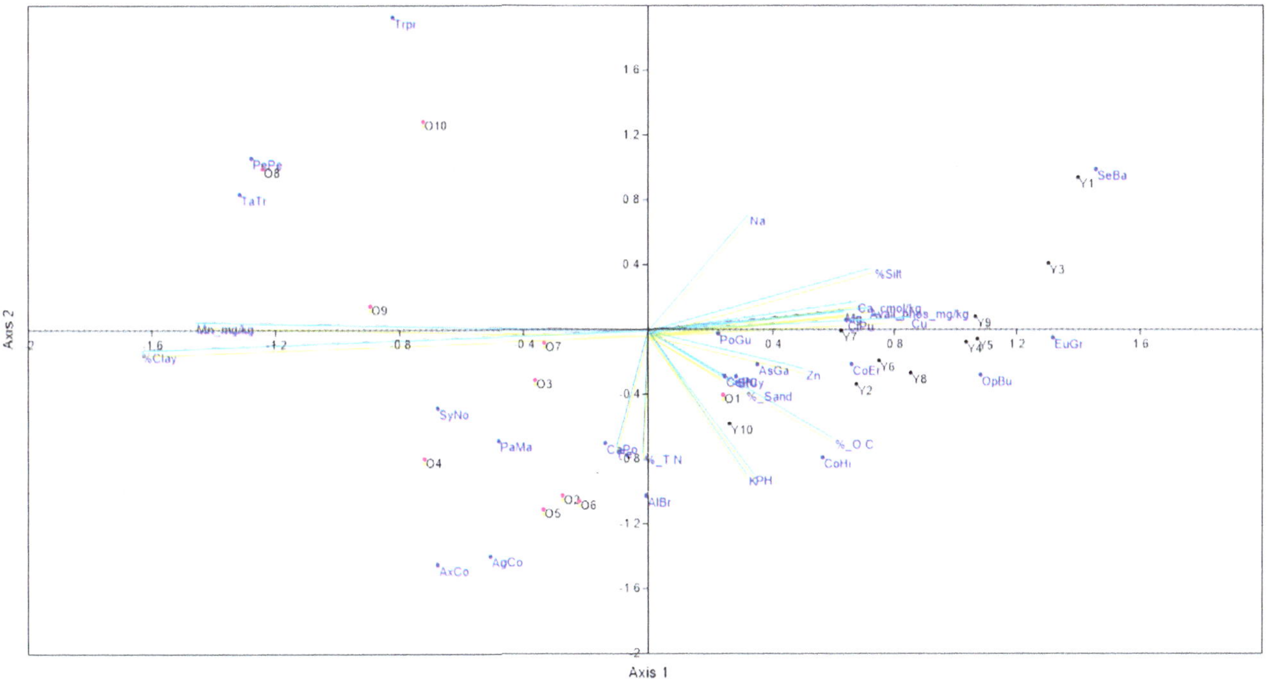

Fig. 1 Canonical correspondence analysis displaying interrelationship between species under orchards of different ages and environmental factors

of the older orchard depicted by O and so the distribution is centered around these points. Also, the place of *Cissus pulpunea* indicated that the species distribution is confined to the young orchard.

6 Conclusion

From the foregoing, we are able to point out the most important soil nutrients factors responsible for plant species distribution of Citrus understory within NIHORT. Further, the research has shown how integrating a temporal dimension into questions that are traditionally spatial can facilitate more rigorous testing of processes we are interested in, as well as bringing about a generation of new, testable hypotheses that would not be evident through the use of spatial approaches alone. This research should be expanded to include other environmental variables such as climate and altitudinal factors. The variability of the soil factors can also be investigated to provide a template useful for informed decision on soil fertilizer application and conservation.

References

Fashae, O., Olusola, A., Faniran, A.: Environmental factors and pattern of riparian vegetation along the downstream section of the lower Ogun River, Nigeria, Singapore. J. Trop. Geogr. **39**(2), 215–223 (2018)

Iloeje, N.P.: A new geography of Nigeria. New rev. ed., Longman Publishers, Ibadan, Nigeria (2001)

Meddad-Hamza, A., Hamza, N., Neffar, S., Beddiar, A., Gianinazzi, S., Chenchouni, H.: Spatiotemporal variation of arbuscular mycorrhizal fungal colonization in olive (*Olea europaea* L.) roots across a broad mesic-xeric climatic gradient in North Africa. Sci. Total Environ. **583C**, 176–189 (2017)

Neffar, S., Chenchouni, H., Beddiar, A., Redjel, N.: Rehabilitation of degraded rangeland in drylands by prickly pear (*Opuntia ficus-indica* L.) plantations: effect on soil and spontaneous vegetation. Ecologia Balkanica **5**, 63–76 (2013)

Ter Braak, C.J.: Canonical community ordination. Part 1: basic theory and linear methods. Ecoscience **1**(2), 127–140 (1994)

Part IV

Investigations and Applications in Environmental Biotechnology

Effect of Climatic Factors on Essential Oil Accumulation in Two Lamiaceae Species from Algerian Semiarid Lands

Souad Mehalaine and Haroun Chenchouni

Abstract

Plant metabolism is vastly influenced by climatic factors in terms of quantity and quality. In this regard, the objective of this study was to assess the impact of some climatic factors on the amount of essential oils (EOs) in two plant species growing spontaneously in an Algerian semiarid zone: *Thymus algeriensis* Boiss & Reut. and *Rosmarinus officinalis* L. for five successive years (2010–2014). Essential oils were extracted from plant aerial parts by hydrodistillation using Clevenger type system and the oil yield rates were estimated in relation to the dry matter of the plants. The obtained results showed significant differences in EO yields during the five years, among the species and the interaction 'Year × Species' ($P < 0.001$). *R. officinalis* accumulated the best oil content during 2010, 2012 and 2013 with 1, 0.93 and 0.88% respectively. *T. algeriensis* yielded higher EOs in 2013 followed by 2012 and 2011 with 1.08, 0.67 and 0.59% respectively. Furthermore, *R. officinalis* produced higher EO amount than *T. algeriensis*. Oils accumulation increased significantly with the increase of precipitation, wind speed and hygrometry but was negatively influenced by air humidity and aridity in both species.

Keywords

Rosmarinus officinalis • *Thymus algeriensis*
Essential oils • Climate effects

S. Mehalaine (✉) · H. Chenchouni
Department of Natural and Life Sciences, Faculty of Exact Sciences and Natural and Life Sciences, University of Tebessa, 12002 Tebessa, Algeria
e-mail: souadmehalaine@yahoo.fr

H. Chenchouni
e-mail: chenchouni@gmail.com

H. Chenchouni
Laboratory of Natural Resources and Management of Sensitive Environments 'RNAMS', University of Oum-El-Bouaghi, 04000 Oum-El-Bouaghi, Algeria

1 Introduction

Volatile oils are the most important odorous principles found in various plants. They play important roles since they are used for their therapeutic action, in pharmaceuticals, foods, cosmetics and perfumery industries. The preparation of volatile oils has been developed in modern times into a large industry (Balbaa et al. 1981). Various factors influence the composition and concentration of secondary metabolites (SMs) in plants, these include (i) habitat ecological conditions, mainly climatic and edaphic factors, that control the plant physiological and metabolism activities and also inter- and intraspecific biological interactions, (ii) geographic and genetic variations (Figueiredo et al. 2008; Daniel 2006). Many species of Lamiaceae family are well known for producing high yields of essential oils (EOs) (Franchomme 1999), among them two aromatic medicinal species were investigated in the current study, viz. *Thymus algeriensis* Boiss & Reut., an endemic species to North Africa and *Rosmarinus officinalis* L. that is common in the Mediterranean region and widely distributed in Algeria (Quezel and Santa 1963). Within this context, the purpose of this research was to estimate the amount of volatile oils in the target species during five years and to attempt to give more interpretations for the effect of some climatic factors on the accumulation and annual variations of these SMs.

2 Materials and Methods

2.1 Plant Collection and Screening

The fresh aerial parts (leaves, stems and flowers) of *Thymus algeriensis* and *Rosmarinus officinalis* were harvested at the flowering phase (March–April) from Ain Beida (35.805°N, 7.376°E, altitude: 920 m a.s.l.) in a semiarid zone of Algeria during the years 2010–2014. The wooden parts of the plants were removed, then the remaining samples were left to dry

H. Chenchouni et al. (eds.), *Exploring the Nexus of Geoecology, Geography, Geoarcheology and Geotourism: Advances and Applications for Sustainable Development in Environmental Sciences and Agroforestry Research*, Advances in Science, Technology & Innovation, https://doi.org/10.1007/978-3-030-01683-8_12

under room temperature until they reached a constant weight. The oils were extracted from 50 g of dry aerial parts by hydrodistillation using Clevenger type apparatus for 2 h following the method of European Pharmacopoeia (2004). The obtained EOs were completely separated from water and the yield was calculated in relation to dry weight of plant aerial parts in three replications.

2.2 Data Management and Statistical Analysis

The EOs variation between the study period and plant species were statistically analyzed using two-way ANOVA, where the interaction of both factors was included. Multiple comparisons of means via Tukey's post hoc test were conducted among the study years separately for each species. The effects of climate conditions of the pre-flowering period (~two months) on the yearly variations of EO contents in both species were investigated using a generalized linear model (GLM) with Gaussian distribution error and identity link. For each year, daily meteorological data of two months preceding plant sample collection were used to compute accumulated precipitation (*PP*) and averages of mean temperatures (*T*), hygrometry (=*PP*/*T*), air humidity, wind speed, and monthly De Martonne's aridity index (=12 *PP*/(*T* + 10)). The statistical analysis was carried out using the R software.

3 Results

3.1 Yearly and Specific Essential Oil Yields

Essential oil yields varied significantly between the five study years (ANOVA: $F_{(4,20)}$= 50.05, $P < 0.001$), the two plant species ($F_{(1,20)}$= 24.71, $P < 0.001$) and the interaction 'Year × Species' ($F_{(4,20)}$ = 27.34, $P < 0.001$). The highest EO accumulation in *Rosmarinus officinalis* was observed during 2010, 2012 and 2013 with 1, 0.93 and 0.88% respectively. However, this species accumulated less EOs during 2011 and 2014 with 0.63 and 0.46%, respectively. *Thymus algeriensis* yielded higher EOs in 2013, than in 2012 and 2011 with 1.08, 0.67 and 0.59%, respectively. However, EO concentrations observed in 2010 and 2014 were significantly the lowest with 0.47 and 0.49%, respectively (Fig. 1).

3.2 Climatic Effects on Essential Oil Yields

The GLM revealed that EO contents varied significantly between medicinal plant species (GLM: $F = 12.12$, $P = 0.002$), where EO yields were higher in *R. officinalis* compared to *T. algeriensis* (Fig. 2, Appendix 1). Essential oils increased significantly with the increase of precipitation ($P = 0.002$), wind speed ($P = 0.024$) and hygrometry ($P = 0.001$). However, EO yields of both species were negatively affected by climatic factors including air humidity ($P = 0.002$) and aridity ($P = 0.001$). The GLM indicated that EOs were deemed positively related to air temperature, but with no significant effect ($P = 0.793$) (Appendix 1).

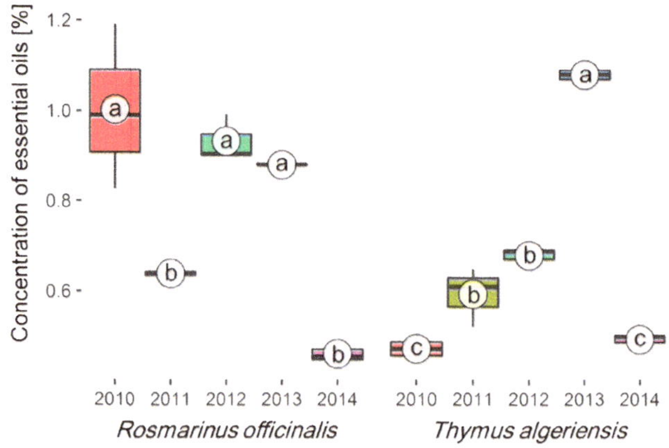

Fig. 1 Box plots displaying yearly variations of essential oil concentrations in *Rosmarinus officinalis* and *Thymus algeriensis* growing in semiarid regions of Algeria. The same letters associated with average values (white circles) are not significantly different following Tukey's post hoc test

4 Discussion

The results indicate that the values of oil amounts for the two species are in accordance with literature. Dob et al. (2006) obtained an oil content of 1.13% on aerial parts from Algerian *T. algeriensis*. However, Amarti et al. (2010) reported a low EO yield (0.3%) from aerial parts of *T. algeriensis* from Morocco. According to Zaouali et al. (2013), the greatest EO yield was obtained at the flowering phase from leaves and stems of Tunisian *R. officinalis* (0.71%). In addition, the oil content varies according to the phenological stage. SM and EOs patterns in plants are specie-specific (Kleiber et al. 2017). However, EO concentrations experience high seasonal variability over the year (Nogués et al. 2015). Regarding the effects of climate on the variation of EO concentrations, our results agree with those of Kleiber et al. (2017) and Kainulainen et al. (1992) that reported that EO contents increase under drought stress but are not affected by waterlogging conditions. Kleiber et al. (2017) suggest that SMs play a core function for abiotic stress tolerance. However, Llusià et al. (2006) and Sharma

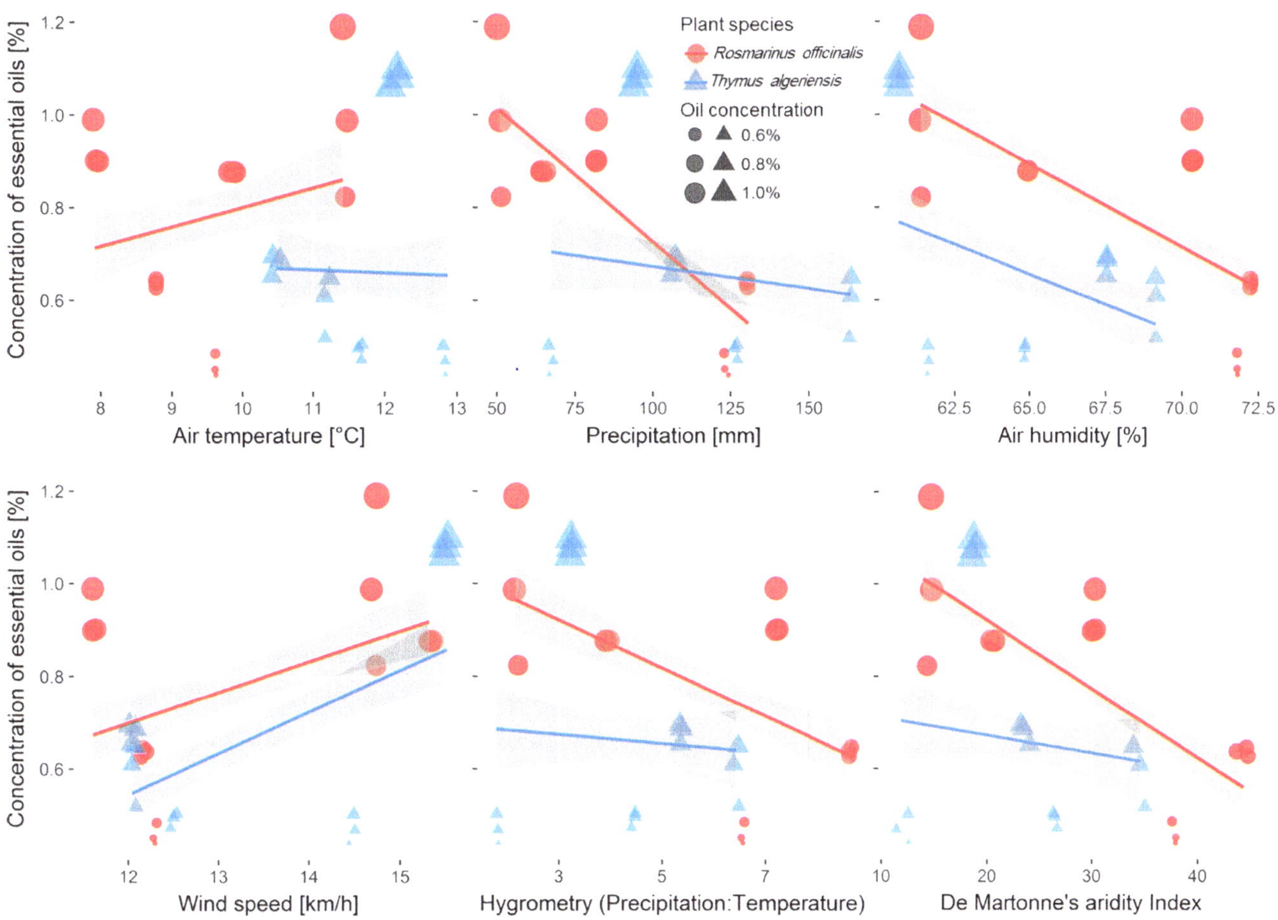

Fig. 2 Effect of climatic factors on the variation of essential oil concentrations in *Rosmarinus officinalis* and *Thymus algeriensis* grown in semiarid regions of Algeria. The lines represent a linear regression with a Gaussian GLM (generalized linear model) fit with confidence intervals in light grey

et al. (2016) showed that SM contents decrease in stressed plants in dry season and/or under greater water and heat stresses, whereas maximum EO concentrations were found in the coldest periods.

5 Conclusion

This research assessed the volatile oil yields in two aromatic plant species: *T. algeriensis* and *R. officinalis* during five years and investigated the effect of some climatic factors on the accumulation of these secondary metabolites. The results showed significant variations of oil content during the different years in both species. Besides, *R. officinalis* accumulated higher EOs compared to *T. algeriensis* and the climatic factors affected significantly the EO production in the tested species. The obtained results can help to understand the plant behavior under semiarid climatic conditions and thus select an appropriate cultivation strategy providing high biomass production and volatile oil accumulation.

Appendix 1

Results from the generalized linear model (Gaussian GLM) testing the effects of climate factors on the variation of essential oil concentrations in *Rosmarinus officinalis* and *Thymus algeriensis* growing in semiarid regions of Algeria.

GLM coefficient summary

Variables	Estimate	*SE*	*t*-value	*P*	Sig.
Intercept	4.209	2.363	1.781	0.089	ns
Thymus algeriensis	−0.627	0.180	−3.481	0.002	**
Mean temperature	0.025	0.093	0.265	0.793	ns
Precipitation	0.010	0.003	3.447	0.002	**
Air humidity	−0.080	0.023	−3.511	0.002	**
Wind speed	0.081	0.034	2.416	0.024	*
Hygrometry	0.400	0.106	3.758	0.001	**
De Martonne aridity	−0.084	0.022	−3.764	0.001	**

Type III F-test

Variables	*Df*	*SS*	*F*	*P*	Sig.
Plant species	1	0.121	12.117	0.002	**
Mean temperature	1	0.001	0.070	0.793	ns
Precipitation	1	0.118	11.882	0.002	**
Air humidity	1	0.123	12.328	0.002	**
Wind speed	1	0.058	5.837	0.024	*
Hygrometry	1	0.141	14.124	0.001	**
De Martonne aridity	1	0.141	14.168	0.001	**
Residuals	22	0.219			

SE: standard error, *P*: probability value, *Df*: degrees of freedom, *SS*: sum squares, *F*: *F*-statistics, Sig.: significance codes.
**: $P < 0.01$, *: $P \leq 0.05$, ^{ns}: $P > 0.05$.

References

Amarti, F., Satrani, B., Ghanmi, M., et al.: Chemical composition and antimicrobial activity of essential oils of *Thymus algeriensis* Boiss & Reut. and *Thymus ciliatus* (Desf.) Benth. from Morocco. Biotechnol. Agron. Soc. Environ. **14**(1), 141–148 (2010)

Balbaa, S.I., Hilal, S.H., Zaki, A.Y.: Medicinal plants constituents. Third ed., Central Agency for University and School Books, Cairo (1981)

Daniel, M.: Medicinal Plants: Chemistry and Properties. CRC Press (2006)

Dob, T., Darhmane, D., Benabdelkader, T., et al.: Studies on the essential oils and antimicrobial activity of *Thymus algeriensis* Boiss & Reut. Inter. J. Aromather. **16**(2), 95–100 (2006)

European Pharmacopoeia, 5th edn. Council of Europe, Strasbourg, France (2004)

Figueiredo, A.C., Barroso, J.G., Pedro, L.G., et al.: Factors affecting secondary metabolite production in plants: volatile components and essential oils. Flavour. Frag. J. **23**(4), 213–226 (2008)

Franchomme, P: L'aromathérapie, Thérapeutique de pointe en médecine naturelle, vol. I-1, 1ère edn. Sinceiro Entreprises Ltd, française (1999)

Kainulainen, P., Oksanen, J., Palomäki, V., et al.: Effect of drought and waterlogging stress on needle monoterpenes of *Picea abies*. Can. J. Bot. **70**(8), 1613–1616 (1992)

Kleiber, A., Duan, Q., Jansen, K., et al.: Drought effects on root and needle terpenoid content of a coastal and an interior Douglas fir provenance. Tree Physiol. **37**(12), 1648–1658 (2017)

Llusià, J., Peñuelas, J., Alessio, G.A., et al.: Seasonal contrasting changes of foliar concentrations of terpenes and other volatile organic compound in four dominant species of a Mediterranean shrubland submitted to a field experimental drought and warming. Physiol. Plant. **127**(4), 632–649 (2006)

Nogués, I., Muzzini, V., Loreto, F., et al.: Drought and soil amendment effects on monoterpene emission in rosemary plants. Sci. Total Environ. **538**, 768–778 (2015)

Quezel, P., Santa, S.: Nouvelle Flore de l'Algérie et des Régions Désertiques Méridionales. Tome 2. Ed. CNRS, Paris (1963)

Sharma, S., Adams, J.P., Sakul, R., et al.: Loblolly pine (*Pinus taeda* L.) essential oil yields affected by environmental and physiological changes. J. Sustain. Fores. **35**(6), 417–430 (2016)

Zaouali, Y., Chograni, H., Trimech, R., et al.: Changes in essential oil composition and phenolic fraction in *Rosmarinus officinalis* L. var. typicus Batt. organs during growth and incidence on the antioxidant activity. Ind. Crops Prod. **43**, 412–419 (2013)

A Biotechnological Method for Breeding Grapes Using in Vitro Growth Stimulants

Abdulmalik Batukaev, Mukail Mukailov, Magomed Batukaev, Tatiana Minkina, and Svetlana Sushkova

Abstract

The aim of the research was to found grape breeding method that allows obtaining virus-free plants during breeding. It used the apical meristem for grapes breeding which is usually free from viruses. The conducted observations showed that a part of meristems was necrotic during the first stage of cultivation (2 weeks). The remaining meristems have evolved into micropiles of 2–2.5 mm which were re-transplanted. One month after planting, the surviving apical meristems were transplanted into a nutrient medium containing the same components. The transplantation was carried out in biological test tubes performed during 45–55 days for formation of regenerants with a size of 6–10 cm and obtaining microclones. The obtained material quality was measured according to growth regulators implemented such as 6-benzylaminopurine (6-BAP) and gibberellic acid (GA). It was found the microshoots grown on a medium with GA and 6-BAP concentration of 0.5 mg/l developed the most effective and virus free grape plants. 40% of the survival rate of the apical meristems was reached which allowed their further cultivation and propagation aiming at the production of virus-free planting material.

Keywords

Grapes • Plant reproduction • Nutrient medium
Growth regulators

A. Batukaev · M. Mukailov · M. Batukaev
Chechen State University, Dudaeva Blvd., 17, 366007 Grozny, Russia

A. Batukaev · M. Batukaev
Chechen Research Institute of Agriculture, Lenina str., 1, village Gikkalo, Chechen Republic Russia, 366021 Grozny, Russia

T. Minkina · S. Sushkova (✉)
Southern Federal University, Bolshaya Sadovaya str., 105, 344006 Rostov-on-Don, Russia
e-mail: snsushkova@sfedu.ru

1 Introduction

Grapes are one of the most common fruit crops, which plays an important role in the world economy. Increasing the production of grapes requires not only the expansion of areas, but also the development and improvement of technologies that ensure the accelerated reproduction of promising varieties. Modern viticulture in Russia should be based on the production of certified planting stock. However, planting material of higher categories in the Russian Federation is not produced (Batukaev 1998).

The main objective of the study was to improve the technology of clonal micropropagation using growth regulators. The task was to obtain a healthy planting stock of grapes and to introduce a certification system for planting material similar to that of European countries.

Modern technology for producing a healthy planting material includes biotechnological methods, integrated treatment with meristem culture, accelerated healthy specimen propagation on artificial nutrient media, and the creation of collections of the improved forms in vitro (Burgutin et al.1983). Clonal micropropagation is the most appropriate example to illustrate how potential of plants (or their individual tissues and organs) can be achieved with the help of biotechnological methods, real reproduction rates being hundreds and even thousands of times higher than in any of the traditional techniques (Vysotskii 2011).

2 Materials and Methods

The object of the study was grape cultivars with complex resistance. Intensively growing green shoots of grapes were used as a parent material. They were cut into single-eye sections and further meristems were isolated in laminar boxes. The experiment was carried out with such cultivars as Augustin, Moldova, Delight, Muscat of Italy, Early Magarach, Gift of Magarach, Viorika (chimeras variety), etc.

H. Chenchouni et al. (eds.), *Exploring the Nexus of Geoecology, Geography, Geoarcheology and Geotourism: Advances and Applications for Sustainable Development in Environmental Sciences and Agroforestry Research*, Advances in Science, Technology & Innovation, https://doi.org/10.1007/978-3-030-01683-8_13

Meristem isolation was followed by sterilization of single-eye cuttings in 2% sodium hypochlorite solution. The sterilized sections were placed in a sterile Petri dish. Before meristems were isolated. The covering scales were sequentially removed from the tip of the eye exposing the apical meristem with primordial leaves. The scales were removed with a dissecting needle under the MBS-10 stereoscopic microscope installed in a dustproof chamber (laminar box). Isolated meristems (200–400 μ) were immediately placed in the agar medium in Petri dishes which in turn were placed in a culture room with the appropriate conditions: illumination 3–4 thousand luces, temperature 27–280 °C, the relative air humidity is 65–70%. The modified nutrient medium MS (Murasige and Skuga) with vitamins was used containing 1 mg/L of thiamin, 1 mg/L of pyridoxine, 1 mg/L of nicotinic acid, 50 mg/L of mesoinositol, 2% of carbon source (sucrose), and 0.7% agar, with pH being adjusted to 6.4–6.5. Various concentrations and combinations of auxins and cytokinins were added to the nutrient medium as growth regulators. The effects produced by indole-butyric acid (IBC) and indole-acetic acid (IAA) belonging to the auxin group, and by 6-benzylaminopurine (6-BAP), 2-isopentyl adenine (2iP), kinetin (cytokinin group), as well as gibberellic acid (GA) were studied. At the first stage, the plant material was grown in Petri dishes then in test tubes (40 × 120 mm) containing 20 mL of a nutrient medium. Explants were transplanted as and when necessary taking into account such indicators as survival of apical meristems and single-eye explants, explant growth intensity, root formation and development.

3 Results and Discussion

The method for obtaining virus-free plants is based on the fact that the content of viruses in the diseased plant decreases towards the shoot tip. The apical meristem is usually virus free. The apical meristem is an apex of actively dividing cells with a height of 0.2–0.4 mm (Vysotskii 2011; Doroshenko 1997; Garre et al. 1979). However, the meristem can be difficult to isolate without damage so one or two leaves primordia were often isolated along with it. A part of meristems (40–60% depending on the variety) were observed to be necrotic during the first stage of cultivation (2 weeks). The remaining meristems have evolved into microshoots of 2–2.5 mm in size a month later after planting. These microshoots were then transplanted into the same nutrient medium in biological test tubes. The survival rates of apical meristems at the stage of introduction into the culture in vitro for the group of table grape varieties such as Augustine, Vostorg, Muscat of Italy, Early Magarach is 50% on average. For technical varieties including Gift of Magarach, Viorica, Rkatsiteli, they had a survival rate of apical meristems of about 40–45%. Apical meristems death at the cultivation stage appears to occur due to damage to the apical structures while dissecting. A month after planting the surviving apical meristems were transferred to a nutrient medium containing the same components. The meristems were transferred to biological test tubes (40 × 120 mm), it took 45–55 days for 6–10 cm long regenerants with a size of two be formed. Then these microplants were divided into cuttings and microclones were obtained. When replanting the shoots of the cluster, their survival rate is high enough, varying from 75% for the Rkatsiteli variety and more than 90% for the Moldovan and Muscat of Italy varieties. The percentage of infected shoots is very low. Apparently, sterilization of apical meristems when introduced into culture in vitro, as well as replanting in sterile conditions (laminar boxes) played a key role. Within 55–60 days plants regenerants 6–10 cm in size were formed, with clonal micropropagation to follow. Plants regenerants were cut into fragments that included a knot with a leaf and a bud (the lower part of the internode is longer than the upper one by 1–2 cm). The obtained microcuttings were planted in biological test tubes (40 × 120 mm) onto agar medium so that the lower part of the internode was immersed in agar. The tubes were covered with foil and placed in a culture room with appropriate conditions. Growth regulators are one of the most important and integral components of the nutrient medium (Gamburg et al. 1990; Kulaeva et al. 1973). The experiments showed that regeneration of shoots from isolated apexes occurred at all concentrations of 6-BAP, except for 5.0 mg/L concentration of the drug when the tips immediately began to stagnate and died. Microshoots grown on a medium with a 6-BAP concentration of 0.1 mg/l developed very slowly. This is probably due to the fact that such low concentrations weakly stimulate plant organogenesis. The concentration range of 6-BAP from 0.5 to 1.0 mg/L was most effective. Nevertheless, it should be noted that 6-BAP concentration of 1.0 mg/L led to the greatest increase in microshoots. The minimum length of a microshoot was observed at 6-BAP concentrations of 0.1 whereas at a concentration of 5.0 mg/l shoot growth was suppressed. This indicates that low concentrations are not sufficient for shoot growth and development. On the contrary, high concentrations inhibit microshoot development (Table 1).

To accelerate the elongation of microshoots the effect of GA in various concentrations in combination with 6-BAP was studied. The experiment has shown that the best result

Table 1 The effect of 6-BAP on development of single-eye cuttings in vitro, microshoot elongation (mm)

Grape cultivars	6-BAP concentrations (mg/L)				LSD
	0.1	0.5	1.0	2.0	
Augustine	4.8 ± 0.8	10.2 ± 3.7	12.1 ± 3.6	8.2 ± 1.5	1.96
Moldova	5.1 ± 1.1	11.5 ± 3.1	11.9 ± 4.2	7.9 ± 2.1	2.06
Vostorg	5.0 ± 1.4	9.9 ± 2.6	10.6 ± 2.5	7.2 ± 1.8	1.84
Italian Muscat	4.8 ± 0.6	9.8 ± 3.3	12.0 ± 4.7	8.1 ± 2.2	2.08
Early Magarach	5.1 ± 1.7	11.6 ± 4.6	11.9 ± 5.1	8.4 ± 1.8	1.86
Gift of Magarach	4.4 ± 0.6	7.0 ± 2.1	9.2 ± 2.3	5.0 ± 1.4	1.56
Viorica	4.6 ± 1.5	8.2 ± 2.7	11.5 ± 3.6	5.8 ± 1.6	2.18
Rkatsiteli	4.9 ± 2.1	9.1 ± 2.4	11.8 ± 3.7	6.1 ± 1.8	2.35

Note: ± means average between 3 replications, LSD: Least significant difference (at P = 0.05)

was achieved when 0.5 mg/L of 6-BAP was combined with 1.0 mg/L of gibberellic acid. The data showed addition of 5.0 mg/L of 6-BAP totally inhibited the microshoot elongation.

4 Conclusion

The survival of the apical meristems from which the plant grows in vitro (10–12 cm) makes it possible to further cultivate and reproduce them (by repeated grafting) that allows obtaining a virus-free planting material. Conditions for the transplantation was carried out in biological test tubes during 45–55 days. The growth regulators used for accelerating of microshoots elongation were used for current research. The most effective doses were 0.5 mg/L of 6-BAP combined with 1.0 mg/L of gibberellic acid.

Acknowledgements The results were obtained in the framework of the RFBR no. 16-35-60051, Ministry of Education and Science of the Russian Federation No. 6.6222.2017/8.9, President of Russian Federation no. MK-3476.2017.5.

References

Batukaev A.A.: Advances in technology of accelerated reproduction and improving the grape plant material by in vitro method, pp. 1–222. Publishing house of Moscow Timiryazev Agricultural Academy, Moscow (1998). (In Russian)

Burgutin A.B., Butenko R.G., Kataeva N.V., Golodriga P.Y.: Rapid clonal propagation of a grape plant. Agric. Biol. **7**, 48–50 (1983). (In Russian)

Doroshenko, N.P.: Increase in the regenerative capacity of meristems in obtaining the virus-free grape material. Grapes Wine Russia **2**, 6–9 (1997). (In Russian)

Garre, M., Martin-Tanguy, J., Mussillon, P.: La cultur de meristemes et la multilpication Végetative in Vitro au service de la pepiniere. Bull. Petits Fruit **14**, 7–65 (1979)

Gamburg, K.Z., Rekoslavskaya, N.I., Shevtsov, S.G.: Auxins in Plant Tissue and Cell Culture, pp. 243. Novossibirsk, Nauka (1990). (In Russian)

Kulaeva, O.N.: Cytokinins, Their Structure and Functions, pp. 264. Moscow, Nauka (1973). (In Russian)

Vysotskii, V.A.: Biotechnological methods in modern gardening, vol. 26, pp. 3–10. Collection of research papers of All-Russian Research Institute of Horticultural Crops Selection, Moscow (2011). (In Russian)

Effects of Inoculation with Rhizospheric *Pseudomonas* on Physiological Responses in the Broad Bean (*Vicia Faba*) Grown Under Copper Stress

Boubaker Idder, Rachid Djibaoui, Abdelhakim Reguieg Yassaad El Hocine, and Abdelhak Djoudi

Abstract

The accumulations of chemical pesticides residues in soils represent one of the greatest risks for the environment and especially for plants. The copper bioviability in rhizosphere is the clearest aspect of this pollution process, causing high levels of metal stress, reducing crops yield by altering physiological and biochemical processes in plants. To face such risks, researchers have frequently studied the effect of plant growth promoting rhizobacteria (PGPR) as enhancer of the plant growth under abiotic stress. The goal of this study was firstly to know the level of broad bean tolerance to metal stress of copper, and its physiological effects in the plant, on the one hand, and the role of plant-*Pseudomonas* interaction in enhancing the plant tolerance level to metal stress, on the other. In this study, three *Pseudomonas* strains were isolated and screened by salinity and copper tolerance, then they were individually used as an inoculum in rhizosphere of broad bean *Vicia faba* (OTONO variety) in the presence of 0; 2.5; 10 and 20 mM.L^{-1} of $CuSO_4$. According to the obtained results, under copper stress conditions with and without bacterial inoculation, the production of biomass and total chlorophyll content were significantly decreased. Copper treatments increased proline content in inoculated plants with P7 and P15 strains and in those which were not inoculated. However, this content was decreased in inoculated plants with P1 strain. The inoculation with P1, P7 and P15 strains motivated the production of fresh biomass and accumulation of proline, and otherwise decreased total chlorophyll content in plants.

B. Idder (✉) · R. Djibaoui · A. Djoudi
Laboratory of Microbiology and Plant Biology, University of ABDELHAMID Ibn Badis, 27000 Mstaganem, Algeria
e-mail: adenib983@gmail.com

A. Reguieg Yassaad El Hocine
Laboratory of Biodiversity and Conservation of Soil and Water, University of ABDELHAMID Ibn Badis, 27000 Mostaganem, Algeria

Keywords

Copper stress • *Pseudomonas* • PGPR • *Vicia faba* • Proline • Chlorophyll

1 Introduction

Heavy metals considerably affect the physiological and biochemical processes of the plant, metal stress reduces also crop productivity. Researchers have frequently studied the effect of plant growth promoting bacteria (PGPR) as promoter under abiotic stress. In this study, the goal was firstly to know the level of broad bean tolerance to stress induced by metal stress of copper in soil, and manifested responses in the plant. The second purpose was to find out the role of plant-*Pseudomonas* interaction in improving the tolerance level of plant to metal stress of copper.

2 Materials and Methods

After preparing the bean seeds for germination and the inoculation solution from the tree screened strains, the first inoculation was applied simultaneously with the transplantation of the germinated seeds by adding 120 mL of the bacterial suspension to each pot. The second inoculation was carried out four weeks after transplantation. After one week of transplanting, the pots were regularly irrigated by the HOAGLAND solution. From the fifth week of transplantation, the irrigation solutions (HOAGLAND solution) are provided by $CuSO_4$ treatments during three weeks, at different doeses: 2.5 mM.L^{-1}, 10 mM.L^{-1} and 20 mM.L^{-1}.

The plantation was carried out in a greenhouse at the Agronomy Workshop in Mazzagran, Abdelhamid Ibn Badis University in Mostaganem located at latitude: 35° 53′31.9″ N; longitude: 0° 5′7.4″E with an average temperature of 28 °

H. Chenchouni et al. (eds.), *Exploring the Nexus of Geoecology, Geography, Geoarcheology and Geotourism: Advances and Applications for Sustainable Development in Environmental Sciences and Agroforestry Research*, Advances in Science, Technology & Innovation, https://doi.org/10.1007/978-3-030-01683-8_14

C in the day and 23 °C at night, and a hygroscopy of 55–75%. The experiment was carried out with 16 pots with 4 seeds per pot and 4 pots for each inoculation treatment. The used bean seeds were from Otono variety. They were germinated after their disinfection with a 25% sodium hypochlorite solution for 15 min and then transplanted into pots of 5 kg of mineralized and sterile sand.

The experiment was carried in a completely randomized design with three replications. The data analysis ANOVA was carried out by STAT BOX 6.40. Means were compared and tested for significance using Student–New mean-Keuls test, at 0.05 level of probability.

3 Results and Discussion

The results show that 2.5, 10 and 20 Mm copper treatments affected proportionally the fresh weight of inoculated and non-inoculated *Vicia faba* plants. Lewis et al. (2001) reported that the excess of copper in the soil plays a toxic role and induces damage to the plant. This leads to growth retardation and leaf chlorosis. The total weight of plant biomass decreases in the presence of copper (Oliveira et al. 2014).

However, the inoculation with P1, P7 and P15 strains increased significantly the fresh weight of inoculated plants in the presence as in absence of copper treatment comparatively with the non-inoculated plants. This result is similar to those of some authors who stated that heavy metal-resistant bacteria isolated from a rhizosphere can be used to support plant growth and increase the accumulation of heavy metals (Andreazza et al. 2010; Chen et al. 2008; Dell'Amico et al. 2008; Kumar et al. 2008; Sheng et al. 2008; Ma et al. 2009). This is due to the bioavailability of copper in soil in the presence of bacterial inoculum (Oliveira et al. 2014) (Fig. 1).

3.1 Total Chlorophyll Content

The results obtained, in the absence of bacterial inoculation, show that the total chlorophyll in the leaves of plants is proportional to the copper concentration, indicating the role of copper in the synthesis of photosynthetic pigments. These results have previously been confirmed by many studies. Chugh and Sawhney (1999) suggested that low concentrations of copper increased the chlorophyll content (a) in the plant of *Elsholtzia splendens*, while this content was reduced by high concentrations. But the chlorophyll (b) content is less influenced by high or low concentrations of Cu (Li et al. 2003). Copper can increase or decrease leaf chlorophyll and alter chlorophyll fluorescence and photosystem II activity (Cook et al. 1997; Pätsikkä et al. 2002).

However, low levels of total chlorophyll content were caused by P1, P7 and P15 strains inoculation, in the presence and absence of copper treatments. This decrease is probably due to the hyper-accumulation of copper affecting chlorophyll. This hyperaccumulation of copper is caused by bacterial inoculation, increasing the bioavailability of copper in the soil. Several previous studies have suggested this effect of bacteria on the bioavailability and bioaugmentation of metals via the production of siderophores considered as chelating agents of metals (Kumar et al. 2008; Sheng et al. 2008; Egamberdiyeva 2007) (Fig. 2).

3.2 Proline Content

The results of proline contents in non-inoculated plants show a significant increase in the proline induced by the copper treatment (10 and 20 mM.L^{-1}), indicating the effect of copper stress. This observation has been reported by several studies on different plant species. High level of copper may be one of the stressors and may induce stress accumulation in plant

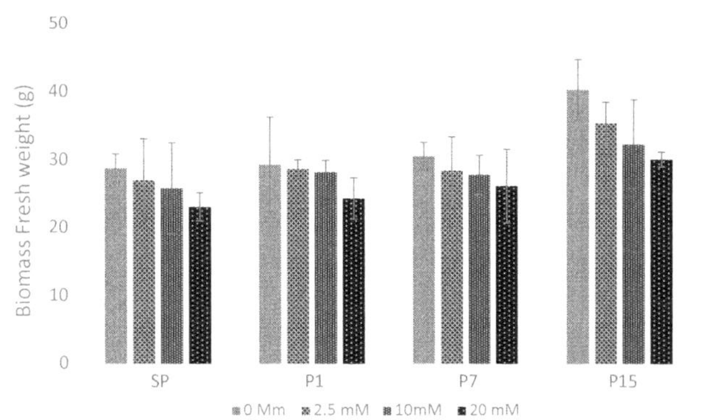

Fig. 1 Combined effect of copper and inoculum on Biomass fresh weight of plant; SP: plant without inoculation; P1, P7 and P15: Treatment with inoculum strains; 0, 2.5, 10 and 20 mM: Copper treatment

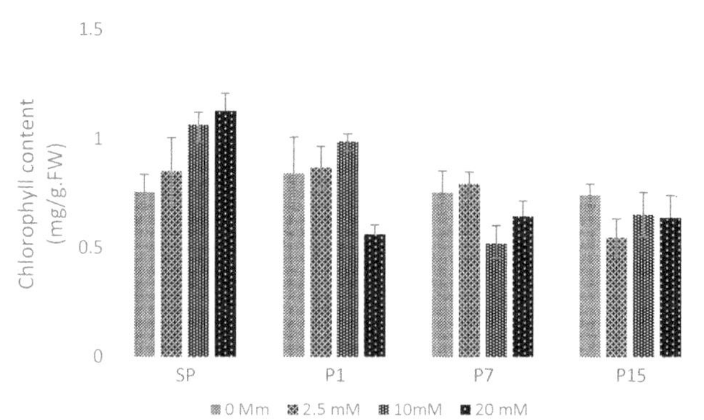

Fig. 2 Combined effect of copper and inoculum on total chlorophyll content in aerial part; SP: plant without inoculation; P1, P7 and P15: Treatment with inoculum strains; 0, 2.5, 10 and 20 mM: Copper treatment

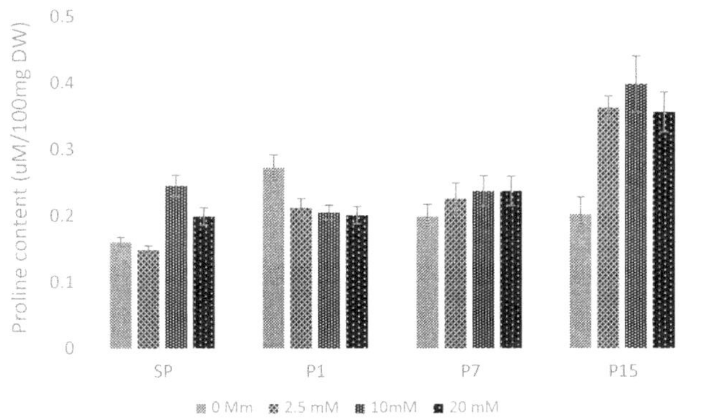

Fig. 3 Combined effect of copper and inoculum on the proline content; SP: plant without inoculation; P1, P7 and P15: Treatment with inoculum strains; 0, 2.5, 10 and 20 mM: Copper treatment

tissues (Oves et al. 2013; Zhang et al. 2008; Joseph et al. 2007; Chen et al. 2001; Miller et al. 2005; Kavi et al. 2005; Alia 2008; Alia and Matysik 2001; Ibrahim et al. 2011). Excess of CuSO4 induced proline accumulation in *C. reinhardtii* (Zhang et al. 2008), and sunflower (Zengin and Kirbag 2007).

However, in the presence of inoculation with P7 and P15, the copper treatments caused a significant increase in proline. This is probably due to the hyper-accumulated copper content in the plant tissues, as shown in many similar studies (Andreazza et al. 2010; Kumar et al. 2008; Sheng et al. 2008). This might be also due to the inoculation strains by stimulation of proline production in the plant organs (Zarea et al. 2012). This effect is significantly clear in inoculated plants with P1. and P15 and were not treated with copper (Fig. 3; Table 1).

4 Conclusions

The results showed that applied copper treatments affected proportionally the decrease of *Vicia faba* biomass, however, the P1, P7 and P15 inoculum has increased the plants biomass. An increase of total chlorophyll content in non-inoculated *Vicia faba* plants was shown; however, the levels of total chlorophyll in the presence of inoculation have significantly decreased. We also found that the effect of copper on proline was distinguished among the inoculation strains, whereas, all the inoculation treatments caused an increase in proline comparatively with non-inoculated plants.

Table 1 Combined effect of copper and inoculum on biomass fresh weight, proline ant chlorophyll content

Bacterial inoculum	Copper treatment (Mm)	Fresh weight (g)	Proline content (uM/100mg DW)		Chlorophyll content (mg/g FW)	
SP	0	28.908 ± 2.107	0.162 ± 0.007	e	0.761 ± 0.081	cdef
SP	2.5	27.103 ± 6.105	0.150 ± 0.006	e	0.857 ± 0.154	bcd
SP	10	25.898 ± 6.721	0.246 ± 0.016	cd	1.072 ± 0.056	a
SP	20	23.193 ± 2.048	0.200 ± 0.013	d	1.135 ± 0.08	a
P1	0	29.443 ± 6.929	0.273 ± 0.02	c	0.847 ± 0.168	bcdef
P1	2.5	28.805 ± 1.361	0.213 ± 0.014	d	0.873 ± 0.099	bc
P1	10	28.335 ± 1.748	0.206 ± 0.011	d	0.995 ± 0.034	ab
P1	20	24.42 ± 3.011	0.202 ± 0.013	d	0.569 ± 0.043	hg
P7	0	30.658 ± 2.036	0.200 ± 0.018	d	0.76 ± 0.098	cdef
P7	2.5	28.547 ± 4.953	0.227 ± 0.023	d	0.799 ± 0.055	cde
P7	10	27.895 ± 2.928	0.238 ± 0.023	cd	0.527 ± 0.082	g
P7	20	26.27 ± 5.412	0.238 ± 0.022	cd	0.652 ± 0.07	defg
P15	0	40.418 ± 4.398	0.203 ± 0.026	d	0.748 ± 0.05	cdef
P15	2.5	35.497 ± 3.093	0.363 ± 0.018	b	0.554 ± 0.086	g
P15	10	32.418 ± 6.524	0.399 ± 0.042	a	0.658 ± 0.103	defg
P15	20	30.202 ± 1.105	0.357 ± 0.03	b	0.644 ± 0.102	efg

Value Average ± standard deviation; Lowercase letter express groups of compared means using the Newman—Keuls test with the value of $P = 0.05$

References

Alia, M.P., Matysik, J.: Effect of proline on the production of singlet oxygen. Amino Acids **21**, 191–203 (2001)

Andreazza, R., Pieniz, S., Okeke, B.C., Camargo, F.A.O.: Evaluation of copper resistant bacteria from vineyard soils and mining waste for copper biosorption. Braz. J. Microbiol. (BJM-1469) (2010)

Chen, W.N., Wu, C.H., James, E.K., Chang, J.S.: Metal biosorption capability of *Cupriavidus taiwanensis* and its effects on heavy metal removal by nodulated *Mimosa pudica*. J. Hazard. Mater. **151**, 364–371 (2008)

Chen, C.T., Chen, L.M., Lin, C.C., Kao, C.H.: Regulation of proline accumulation in detached rice leaves exposed to excess copper. Plant Sci. **160**, 283–290 (2001)

Chugh, L.K., Sawhney, S.K.: Photosynthetic activities of *Pisum sativum* seedlings grown in presence of cadmium. Plant Physiol. Biochem. **37**, 297–303 (1999)

Cook, C.M., Kostidou, A., Vardaka, E., Lanaras, T.: Effects of copper on the growth, photosynthesis and nutrient concentrations of Phaseolus plants. Photosynthetica **34**(2), 179–193 (1997)

Dell'Amico, E., Cavalca, L., Andreoni, V.: Improvement of *Brassica napus* growth under cadmium stress by cadmium resistant rhizobacteria. Soil Biol. Biochem. **40**, 74–84 (2008)

Egamberdiyeva, D.: The effect of plant growth promoting bacteria on growth and utrient uptake of maize in two different soils. Appl. Soil. Ecol. **36**, 184–189 (2007)

Ibrahim, A.S.S., El-Tayeb, M.A., Elbadawi, Y.B., Al-Salamah, A.A.: Bioreduction of Cr (vi) by potent novel chromate resistant alkaliphilic *Bacillus* sp. Strain KSUCr5 isolated from hypersaline soda lakes. Afric. J. Biotechnol. **10**, 7207–7218 (2011)

Kavi, K.P.B., Sangam, S., Amrutha, R.N., Laxmi, P.S., Naidu, K.R., Rao, K.R.S., Rao, S., Reddy, K.J., Theriappan, P., Sreenivasulu, N.: Regulation of proline biosynthesis, degradation, uptake and transport in higher plants: its implications in plant growth and abiotic stress tolerance. Curr. Sci. **88**, 424–438 (2005)

Kumar, K.V., Singh, N., Behl, H.M., Srivastava, S.: Influence of plant growth promoting bacteria and its mutant on heavy metal toxicity in *Brassica juncea* grown in fly ash amended soil. Chemosphere **72**, 678–683 (2008)

Lewis, S., Donkin, M.E., Depledge, M.H.: Hsp70 expression in *Enteromorpha intestinalis* (Chlorophyta) exposed to environmental stressors. Aquat. Toxicol. **51**, 277 (2001)

Li, Y.M., Chaney, R., Brewer, E., Rosenberg, R., Angle, S.J., Baker, A. J.M., Reeves, R.D.: Nelkin J Development of technology for commercial phytoextraction of nickel: economic and technical considerations. Plant Soil **249**, 107–115 (2003)

Ma, Y., Rajkumar, M., Freitas, H.: Inoculation of plant growth promoting bacterium *Achromobacter xylosoxidans* strain Ax10 for the improvement of copper phytoextraction by *Brassica juncea*. J. Environ. Manage. **90**, 831–837 (2009)

Miller, G., Stein, H., Honing, A., Kapulnik, Y., Zilberstein, A.: Responsive modes of *Medicago sativa* proline dehdrogenase genes during salt and recovery dictate free proline accumulation. Planta **222**, 70–79 (2005)

Oliveira, T., Mucha, A.P., Reis, I., Rodrigues, P., Gomes, C.R., Almeida, C.M.R.: Copper phytoremediation by a salt marsh plant (*Phragmites australis*) enhanced by autochthonous bioaugmentation. Mar. Pollut. Bull. https://doi.org/10.1016/j.marpolbul.2014.08.038 (2014)

Oves, M., Khan, M.S., Zaidi., A.: Chromium reducing and plant growth promoting novel strain *Pseudomonas aeruginosa* OSG41 enhance chickpea growth in chromium amended soils. Eur. J. Soil Biol. **56**:72–83 (2013)

Pätsikkä, E., Kairavuo, M., Sersen, F., Aro, E.-M., Tyystjärvi, E.: Excess copper predisposes photosystem II to photoinhibition in vivo by outcompeting iron and causing decrease in leaf chlorophyll. Plant Physiol. **129**(3), 1359–1367 (2002)

Sheng, M., Tang, M., Chen, H., Yang, B., Zhang, F., Huang, Y.: Influence of arbuscular mycorrhizae on photosynthesis and water status of maize plants under salt stress. Mycorrhiza **18**, 287–296 (2008)

Zarea, M.J., Hajinia, S., Karimi, N., Mohammadi, Goltapeh E., Rejali, F., Varma, A.: Effect of *Piriformospora indica* and *Azospirillum* strains from saline or non-saline soil on mitigation of the effects of NaCl. Soil Biol. Biochem. **45**, 139–146 (2012)

Zengin, F.K., Kirbag, S.: Effects of copper on chlorophyll, proline, protein and abscisic acid level of sunflower (*Helianthus Annuus* L.) seedlings. J. Environ. Biol. **28**(3), 561–566 (2007)

Zhang, L.P., Mehta, S.K., Liu, Z.P., Yang, A.Z.: Copper-induced proline synthesis is associated with nitric oxide generation in *Chlamydomonas reinhardtii*. Plant Cell Physiol. **49**(3), 411–419 (2008)

Part V

Spatiotemporal Patterns of Marine Biodiversity and Terrestrial Paleobiodiversity

Arcellinida (Testate Lobose Amoebae) as Sensitive Bioindicators of Arsenic Contamination in Lakes

R. Timothy Patterson, Nawaf A. Nasser, Braden R. B. Gregory, Melody J. Gavel, Etienne Menard, Jennifer M. Galloway, and Helen M. Roe

Abstract

Arcellinida (testate lobose amoebae) are sensitive to arsenic (As) contamination from historic gold mining in Canadian subarctic lakes. Partial redundancy analysis revealed that As (9.4%) and S1 (labile organic matter, 8.9%) were the most important contributors to variance in faunal distribution. Arcellinida are important intermediary food web components in lacustrine environments. Microbes are their primary food source and elevated As-levels are known to suppress microbial growth (S1). The observed correlations are likely to be related to As-destabilization of Arcellinida microbial food supplies. To assess the utility of arcellinidans as tracers of temporal variation in As, we analyzed a freeze core from Frame Lake in the city of Yellowknife, which requires remediation following a steep decline in water quality through the 1960s that left it dead. Land-use activities in the area of the lake are well documented but little information is available regarding their impact on the lake itself. Geochemical, sedimentological and arcellinidan analysis showed that the system was non-depositional from the mid Holocene until ~1962 when the lake began to rapidly infill with highly As contaminated sediments. Since the early 1990s runoff from the catchment, and reduced lake circulation associated with installation of a causeway with rarely opened sluiceways at the only outlet, has caused eutrophication.

R. T. Patterson (✉) · N. A. Nasser · B. R. B. Gregory
M. J. Gavel · E. Menard
Department of Earth Sciences, Ottawa-Carleton Geoscience Center, Carleton University, 1125 Colonel by Drive, Ottawa, ON K1S 5B6, Canada
e-mail: tim.patterson@carleton.ca

J. M. Galloway
Geological Survey of Canada, Calgary, AB T2L 2A7, Canada

H. M. Roe
School of Natural and Built Environment, Queen's University Belfast, Belfast, BT7 1NN, UK

Keywords

Arcellinida • Testate lobose amoebae • Arsenic contamination • Lakes • Subarctic canada

1 Introduction

Due to a strong affinity between gold and As-rich sulfides in mineralized zones, gold mining is a significant source of As contamination in mining districts worldwide. Arsenic contamination of lake systems is of particular concern due to the recreational and aquatic ecosystem services provided by lakes. Lake sediments sequester As, which if physically disturbed, or under altered redox conditions due to climate variability, can remobilize As into the water column causing severe ecological stress and increase human health risks (Galloway et al. 2017). Legacy gold mines (e.g. Giant Mine [GM]; 1948-2004; Fig. 1a, b) near the city of Yellowknife, Northwest Territories, Canada, require considerable remediation of legacy contamination. Mineral processing at GM released As_2O_3 to the atmosphere through roaster stack emissions resulting in extreme contamination of the landscape, including lakes and streams (Galloway et al. 2017). Frame Lake within Yellowknife is of special interest, as it is heavily contaminated due to both urbanization in the lake catchment and mining activities (Nasser et al. 2016). Arcellinida (testate lobose amoebae) are important intermediary food web components in lacustrine environments that rely on bacteria and other microbes as their principle food supply (Patterson et al. 2012). Their agglutinated shells are abundant and preserve well in the sedimentary record making them ideal bioindicators for tracking temporal trophic change. Arcellinida were studied in 90 lakes in the vicinity of the GM to: (1) quantify their sensitivity to As; and (2) use these results to track hydrological changes in Frame Lake resulting from mining and urbanization.

H. Chenchouni et al. (eds.), *Exploring the Nexus of Geoecology, Geography, Geoarcheology and Geotourism: Advances and Applications for Sustainable Development in Environmental Sciences and Agroforestry Research*, Advances in Science, Technology & Innovation, https://doi.org/10.1007/978-3-030-01683-8_15

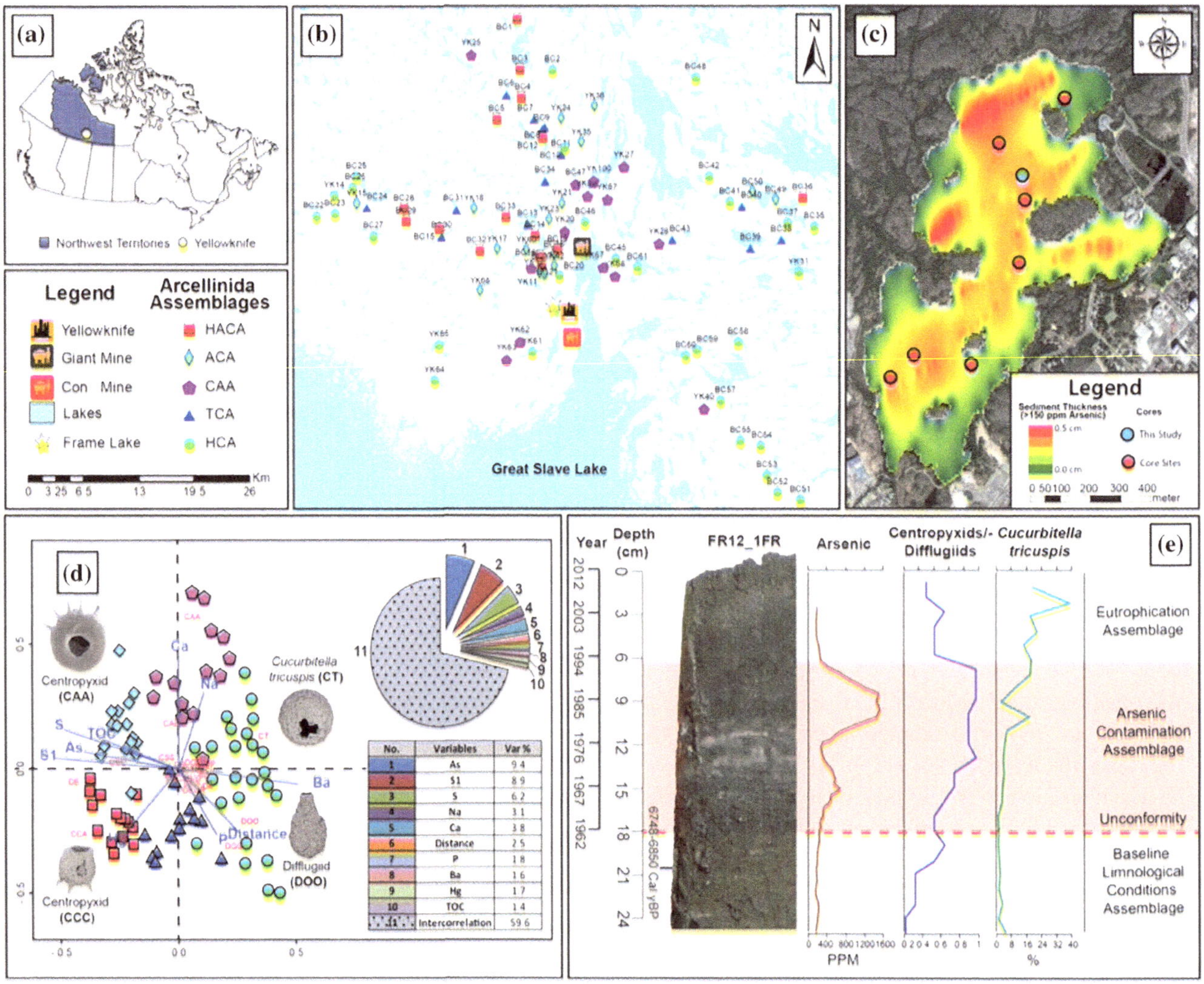

Fig. 1 **a** Map showing Yellowknife, NT. **b** Location of 90 sampled lakes and Frame Lake relative to GM, and distribution of Arcellinida assemblages. **c** Core locations and interpolated thickness of the As-contaminated sediments in Frame Lake. **d** RDA species-environment-sample triplots for sediment-water interface samples having statistically significant values. Partial RDA pie-chart shows variance explained by 10 ecologically significant environmental variables. **e** Paleolimnological reconstruction of Frame Lake Freeze core showing development of As-contaminated conditions post-1962 and the transition to eutrophication conditions in the 1990s

2 Materials and Methods

Ninety-three sediment-water interface samples (SWIS; top 0.5 cm) were collected with a pontoon-equipped helicopter using an Ekman grab from 90 lakes within a ~30 km radius around the site of GM (BC12, YK14; Fig. 1b). Water property data were collected using a YSI Professional Plus multi-parameter instrument at 1-m increments through the water column [pH; temp.; conductivity; oxygen]. A freeze core was collected from Frame Lake and transported frozen to Carleton University. Sub-bottom profiles for Frame Lake were obtained from 14 GPS-positioned transects using a Lowrance HDS-8 sonar and SyQwest Strata Box HD. The profiles were ground-truthed from the stratigraphy observed in 6 gravity cores collected across the lake (Fig. 1c).

The SWIS and Frame Lake samples were characterized by grain size and ICP-MS analysis (Aqua-Regia), Rock Eval (S1), and Arcellinida enumeration. Twenty-five of 30 arcellinidan species and strains observed in 84 SWIS samples had statistically significant populations for Redundancy (RDA) and partial redundancy (pRDA) statistical analysis. The Frame Lake core contained 32 statistically significant species and strains in 30 samples that were statistically significant and retained for subsequent analysis.

3 Results, Discussion and Conclusion

ICP-MS analysis revealed that As concentration is higher than the levels set by the Canadian interim freshwater quality guidelines (5.9 ppm) and probable effect levels (17 ppm) in all samples, particularly in the prevailing downwind area to the west ($\overline{As}$ = 694 ppm ± 1087.3 SD) and north ($\overline{As}$ = 674.5 ppm ± 1956.8 SD) of GM. Conversely, mean As levels in lakes to the east ($\overline{As}$ = 88 ppm ± 136.4 SD) and south ($\overline{As}$ = 76.3 ppm ± 87.5 SD) of the mine site were much lower [cf. 1]. Results of a Q-mode cluster analysis, corroborated by NMDS and RDA (Fig. 1b, d), revealed five distinct arcellinidan assemblages: (1) "High As Contamination Assemblage (HACA)"; (2) "As contamination Assemblage (ACA)"; (3) "*Centropyxis aculeata* Assemblage (CAA)"; (4) "Transitional Conditions Assemblage (TCA)"; and (5) "Healthy Conditions Assemblage (HCA)". The RDA and pRDA analysis revealed that the arcellinidan assemblages reflected ecological conditions characteristic of stressed (HACA, ACA), transitional (CAA, TCA) and relatively healthy lacustrine systems (HCA).

Ten environmental parameters accounted for ~40% of the variance in arcellinidan distribution with As (9.4%) and S1 (labile organic matter 8.9%) being the most influential. There was a strong correlation between the distribution of arcellinidan contamination assemblages and the higher As typical in proximal downwind lakes to the W and NW of the GM (Fig. 1d). The arcellinidan response to the spatial variability of S1 was similar to that observed for As. S1 provides an organic substrate suitable for microbial growth, which in turn mediates the authigenic precipitation of As sulfides (Galloway et al. 2017). Healthy microbial communities thrive within organic substrates with a readily biodegradable labile fraction (i.e. S1), which in turn provides a food supply for grazing Arcellinida. In contrast As is toxic to most bacteria as it can inhibit basic cellular functions linked to energy metabolism, basal respiration, and enzyme activities. The resultant reduction in microbial biomass may induce sufficient environmental stress that nutrition sensitive arcellinidan taxa (e.g. *Cucurbitella tricuspis* (CT), *Difflugia oblonga* "oblonga" DOO) are negatively impacted and stress tolerant taxa such as centropyxids (e.g. *C. aculeata* "aculeata" (CAA); *C. aculeata* "discoides" CAD) take advantage of the open niche space and increase their populations (Fig. 1d). Reflectors observed in the sub-bottom profiles from throughout the Frame Lake and ground-truthed based on ICP-MS results and sedimentology of core data were used to estimate the distribution and volume of legacy arsenic contamination. The geomatics IDW procedure was used to extrapolate the thickness of sediments with >150 ppm As throughout the lake. Large deposits of highly contaminated sediments (As concentrations of 150–1538 ppm) occurred in both basins with up to 57 cm of highly contaminated sediments found along the NW shore (Fig. 1c). The total amount of highly contaminated sediment is estimated at ~230,000 m^3. Down-core arcellinidan and ICP-MS As concentrations were assessed from a frozen core collected in the north basin of Frame Lake as part of an initiative to assess remediation measures required to rejuvenate the lake following significant degradation in the 1960s. The results showed that the system was non-depositional from the mid-Holocene until ~1962 after which the lake began to rapidly infill with highly As contaminated sediments derived from urban development and mine activities. The apparent sharp peak in As in sediments deposited during the mid-1980s is likely the result of remobilization due to changes in redox conditions. Temporal changes in the ratio between the healthy lake indicating difflugiids and As-tolerant centropyxids better characterizes the interval from ~1962 to early 1990s when arsenic significantly impacted the lake ecosystems. From ~1990 onward the major ecological challenge to the lake has been eutrophication derived from runoff and a reduction in lake circulation associated with the construction of a causeway with rarely opened sluiceways at the only outlet.

References

Galloway, J.M., Palmer, M., Swindles, G.T., Jamieson, H.E., Parsons, M.B., Sanei, H., Macumber, A.L., Patterson, R.T., Falck, H.: Organic matter control on the distribution of arsenic in lake sediments impacted by ~65 years of gold ore processing in subarctic Canada. Sci. Total Env. 622–623:1668–1679 (2017)

Nasser, N.A., Patterson, R.T., Roe, H.M., Galloway, J.M., Falck, H., Palmer, M.J., Spence, C., Sanei, H., Macumber, A.L., Neville, L.A.: Lacustrine arcellinina (testate amoebae) as bioindicators of arsenic contamination. Microb. Ecol. **72**, 130–149 (2016)

Patterson, R.T., Roe, H.M., Swindles, G.T.: Development of a thecamoebian (testate amoebae) based transfer function for sedimentary phosphorous in lakes: applications to a record from southern Ontario Canada. Palaeogeography Paleoclimatology Palaeoecology. **348–349**, 32–44 (2012)

Massive Molluscan Shell Accumulation on the Sea Side of Land Strips of Amvrakikos Gulf Lagoon Complex (NW Greece)

Konstantinos Tsolakos, George Katselis, and John A. Theodorou

Abstract

The analysis of the molluscan shells deposition on the sea side of land strips of the lagoon complex of the NW Amvrakikos Gulf shows that bivalve (61.10%) and gasteropod (12.48%) shells constitute the major share of the organogenic material of the sea side land strips. Major representative bivalve species are *Cerastoderma glaucum* (35.32%) and *Polititapes aureus* (17.89%), while a range of other species, such as *Loripes orbiculatus* (4.06%), *Mytilus galloprovincialis* (2.76%), *Peronidia albicans* (2.66%), *Chamelea gallina* (1.80%), *Modiolus barbatus* (1.03%), *Parvicardium exiguum* (1.00%) were also present.

Keywords

Amvrakikos • Lagoon land strips • Molluscan shell accumulation

1 Introduction

Coastal lagoons such as those of Amvrakikos Gulf in W. Greece (Ionian Sea), are highly productive ecosystems, supporting large scale of biodiversity (Ramsar Convention), that are threatened by the environmental conditions and human activities (Katselis et al. 2013; Kountoura and Zacharias 2013; Tsotsios et al. 2016; Theodorou et al. 2018). Sustainable management of this resource is critical and depends on the knowledge of its key ingredients and their function (Rodrigues et al. 2015; Schneider-Storz et al. 2008). Targeting this issue, it is crucial to provide long-term estimations for sandbanks that are at risk of erosion and therefore to take appropriate management measures in due time, ensuring high productivity and lagoons biodiversity as priority habitats in accordance with European Directive 92/43/EEC. The present study focused on the definition of the role of dead mollusks accumulation on the sea side of land strips of the lagoons. Qualitative and quantitative taxa contribution was investigated and discussed below.

2 Materials and Methods

Samples were taken from five stations, along the sea side of the land strips of the coastline of Tsoukalio and Logarou lagoons in the northern part of the Amvrakikos Gulf (Fig. 1a) by placing a 1 m^2 collecting frame and collecting to a depth of 10 cm (Fig. 1b). Empty shells of bivalves, gastropods and other organogenic materials [sieved larger (>) than 2 mm] were separated. Mollusc shells weight was measured. Species identification is based on shell morphology. Bivalve taxa were separated accordingly the left/right valve presence, and the species frequencies were assessed (Schneider-Storz et al. 2008; Poutiers 1987).

K. Tsolakos · G. Katselis · J. A. Theodorou (✉)
Department of Fisheries and Aquaculture Technology, The Western Greece University of Applied Sciences (TEI of Western Greece), Nea Ktiria TEI, 30200 Mesolonghi, Greece
e-mail: jtheo@teimes.gr

H. Chenchouni et al. (eds.), *Exploring the Nexus of Geoecology, Geography, Geoarcheology and Geotourism: Advances and Applications for Sustainable Development in Environmental Sciences and Agroforestry Research*, Advances in Science, Technology & Innovation, https://doi.org/10.1007/978-3-030-01683-8_16

Fig. 1 **a** Sampling stations (St) in the northern part of Amvrakikos Gulf: St.1 to St.3 across the sea side of the land strips of "Tsoukalio" lagoon coastline, and stations St.4 to St.5 across "Logarou" lagoon coastline. **b** Taking samples in station St.3

3 Results

3.1 Qualitative and Quantitative Identification

The total weight of all the collected samples is 4424 g, constituting by 33,837 shells. The major share of organogenic shell mass deposition is accounted for by bivalves (61.10%) and gastropods (12.48%). *Cerastoderma glaucum* (35.32%) and *Polititapes aureus* (17.89%) are the most abundant bivalves, followed by other species, such as *Loripes orbiculatus* (4.06%), *Mytilus galloprovincialis* (2.76%), *Peronidia albicans* (2.66%), *Chamelea gallina* (1.80%), *Modiolus barbatus* (1.03%), *Parvicardium exiguum* (1.00%). Regarding the species distribution, 37 bivalve taxa were recorded in "Logarou" lagoon while 39 were recorded in "Tsoukalio" lagoon (Fig. 2). The major part of total shelly mass weight of *Cerastoderma glaucum* (81.99%) is located in Tsoukalio, while *Polititapes aureus* seems to have similar weight distribution in both lagoons-Tsoukalio (53.76%) and Logarou (46.24%).

4 Discussion

Dead mollusk shells are accumulated due to their stranding on the sea side of the lagoon land strips in N. Amvrakikos gulf, referred to as bivalves and gasteropods with major represenatives *Cerastoderma glaucum* and *Polititapes aureus*. Tsoukalio and Logarou lagoons have similar taxa composition. The role of the species biodiversity as well as the ecosystem services provided by this phenomenon need further investigation. However, an additional taphonomic analysis of mollusk shells could illustrate the deposition of shell aggregates on the sea side of the lagoon land strips as well as the conditions that led to this evidence (Schneider-Storz et al. 2008; Kidwell et al. 2001; Kowalewski et al. 1994).

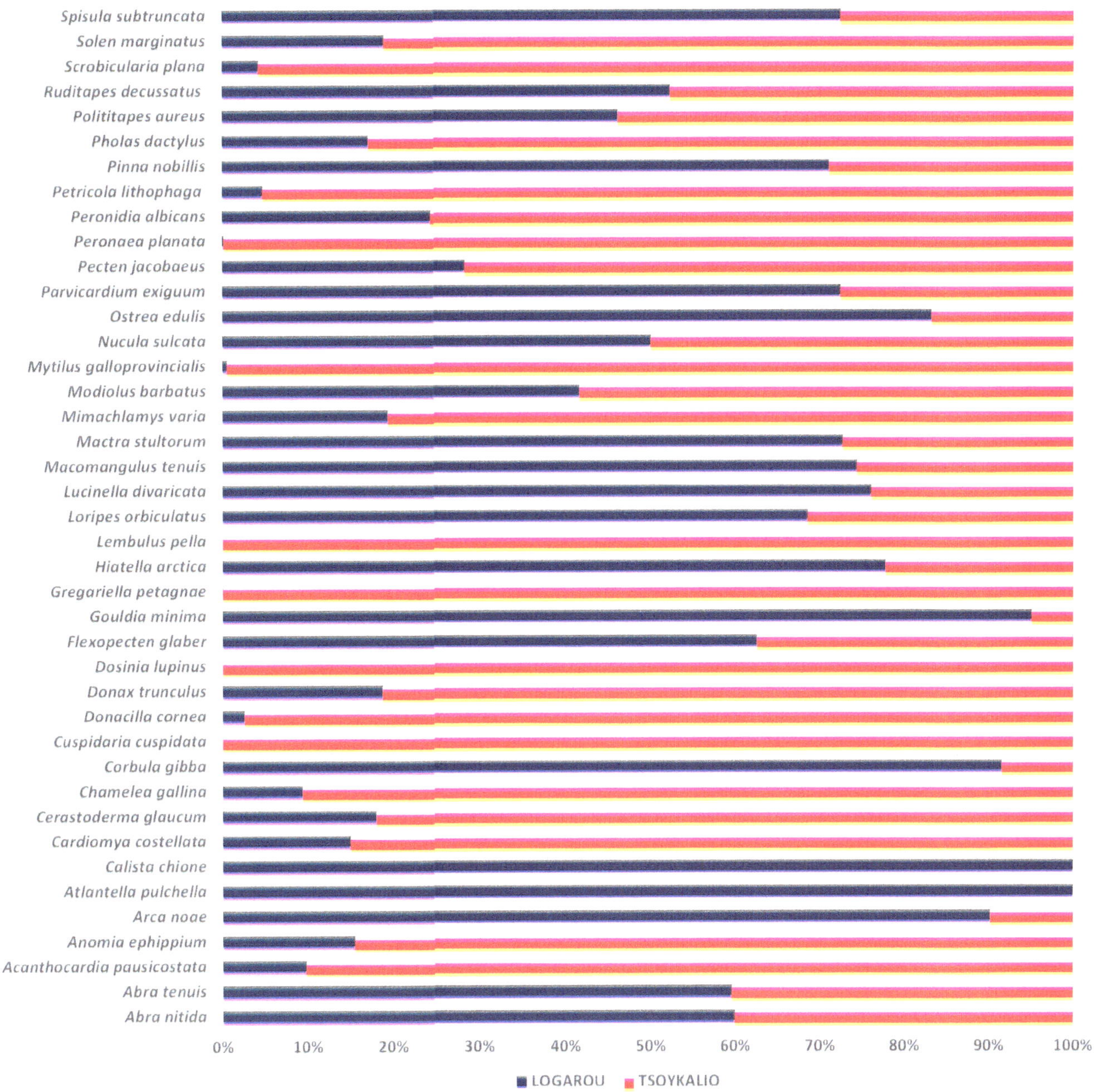

Fig. 2 Species shelly mass weight percentage deposition on the sea side of the land strips of N. Amvrakikos lagoons (Tsoukalio, Logarou)

References

Katselis, G.N., Moutopoulos, D.K., Dimitriou, E.N., Koutsikopoulos, C.: Long-term changes of fisheries landings in enclosed gulf lagoons (Amvrakikos gulf, W Greece): influences of fishing and other human impacts. Estuar. Coast. Shelf Sci. **131**, 31–40 (2013)

Kidwell, S.M., Rothfus, T.A., Best, M.M.: Sensitivity of taphonomic signatures to sample size, sieve size, damage scoring system, and target taxa. Palaios **16**(1), 26–52 (2001)

Kountoura, K., Zacharias, I.: Trophic state and oceanographic conditions of Amvrakikos Gulf: evaluation and monitoring. Desalin. Water Treat. **51**, 2934–2944 (2013)

Kowalewski, M., Flessa, K.W., Aggen, J.A.: Taphofacies analysis of Recent shelly cheniers (beach ridges), northeastern Baja California, Mexico. Facies **31**(1), 209 (1994)

Poutiers, J.M.: Bivalves (Acephales, Lamellibranches, Pelecypodes). In Fischer, W., M. Schneider, M. (eds.) Fiches FAO D' Identification des espèces pour les besoins de la pêche. Méditerranée et Mer Noire zone de pêche 37. Revision 1. Végétaux et Invertébrés, vol. 1, pp. 371–512 (1987)

Rodrigues, L.C., van den Bergh, J.C.J.M., Massa, F., Theodorou, J.A., Ziveri, P., Gazeau, F.: Sensitivity of Mediterranean bivalve mollusc aquaculture to climate change and ocean acidification: results from a producers' survey. J. Shellfish Res. **34**(3), 1161–1176 (2015)

Schneider-Storz, B., Nebelsick, J.H., Wehrmann, A., Federolf, C.M.: Comparative taphonomy of three bivalve species from a mass shell accumulation in the intertidal regime of North Sea tidal flats. Facies **54**(4), 461 (2008)

Theodorou, J.A., Dimitriou, E., Katselis, G.: The crisis management of Amvrakikos Gulf (W. Greece) massive fish mortality: lessons learned from the handling of a 950 tons dead farmed fish biomass. In: Societal and Governing Responses to Global Change in Marine Systems, 1st edn. Routledge Studies in Environment, Culture, and Society Series. New York (2018)

Tsotsios, D., Tzovenis, I., Katselis, G., Geiger, S., Theodorou, J.A.: Spat settlement of the smooth scallop *Flexopecten glaber* (L 1758), and variegated scallop *Chlamys varia* (L1758) in Amvrakikos Gulf, Ionian Sea (NW Greece). J. Shellfish Res. **35**(2), 467–474 (2016)

Vertical Distribution of Recent Benthic Ostracoda at Pullivasal and Kurusadai Islands, Gulf of Mannar, Southeast Coast of India

Kumari Deepali, Sithu G. D. Sridhar, Shaik M. Hussain, and Subramanian Maniyarasan

Abstract

Two core samples were collected during April 2017 from the territorial zone of two islands in the Gulf of Mannar; Pullivasal and Kurusadai for a length of about 1.05 and 0.87 m respectively, at a low tide of the day. Calcium carbonates, organic matter, type of substrate were determined at a resolution of 2.5 cm for both cores. A total of thirty-six Recent Benthic Ostracoda species were identified. The species as well as their assemblage were found and compared with the parameters determined. The vertical distribution of species and their diversity, abundant species and the occurrence of cosmopolitan species within the region of study were discussed. The study is based on the environmental parameters that promotes Ostracoda diversities as well as for their abundance, in order to create an information system on the resources of Gulf of Mannar.

Keywords

Assemblage • Abundant and cosmopolitan Pullivasal and Kurusadai Isands • Recent benthic Ostracoda • Vertical distribution

1 Introduction

Ostracods are a well documented group within the Crustaceans. They are known to inhabit a wide variety of aquatic environments. The areas chosen for investigation, Pullivasal and Kurusadai Islands are part of the tropical Mandapam Islands group of Mannar Gulf. In order to protect the biodiversity of this area, the government declared the Gulf of Mannar a Marine National Park (GOMMBRE 1997). Some notable work has already been achieved (Kathiresan 2006) on some of the islands and research gaps being found. To add new findings, this study was undertaken. The study area is characterized by narrow sandy coastlines with off-shore reefs of dead coral and other calcareous structures on sandy bottoms (Government of India, Department of Ocean Development 2001). As the southern part of these islands show coral ecosystem and the northern part mangroves, the study was undertaken on the territorial zones where Kurusadai represents mangroves and Pullivasal represents corals. The study was carried out to find out vertical distribution of Recent Benthic Ostracoda between these two environments.

2 Materials and Methods

In the study area 1.05 and 0.87 m long sediment cores were collected from the coastal territorial zones of the two islands, Pullivasal and Kurusadai. We compared Ostracod species diversity between the two collection sites, using 2.5 cm interval sub-samples for both cores. Given the difference in length of the cores, we used 42 sub-samples from the core collected from Pullivasal, and 35 sub-samples from Kurusadai Island. $CaCO_3$ (Piper 1947), organic matter (OM) (Gaudette et al. 1974) and type of substrate (Krumbein and Pettijohn 1938) were determined for each sub-sample. We followed the Trefethen classification (Trefethen 1950) for substrate determination. The specimens of benthic Ostracoda were identified from 2.5 g of samples from each subsample of 2.5 cm interval. Figure 1 displays the study area location map.

K. Deepali (✉) · S. G. D. Sridhar · S. Maniyarasan
Department of Applied Geology, University of Madras, 600025 Chennai, India
e-mail: deepaliranigeology@gmail.com

S. M. Hussain
Department of Geology, University of Madras, 600025 Chennai, India

H. Chenchouni et al. (eds.), *Exploring the Nexus of Geoecology, Geography, Geoarcheology and Geotourism: Advances and Applications for Sustainable Development in Environmental Sciences and Agroforestry Research*, Advances in Science, Technology & Innovation, https://doi.org/10.1007/978-3-030-01683-8_17

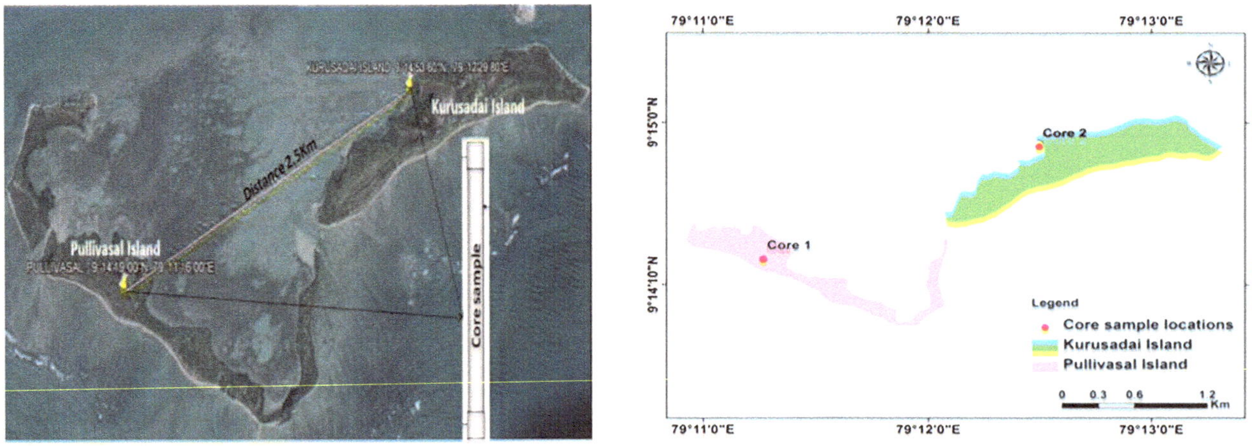

Fig. 1 Location of the used sample in the study

3 Results

The results of sediment parameters, Calcium carbonate, organic matter and type of substrate of the cores collected from Pullivasal (C1) and Kurusadai (C2) Islands for each depth interval with corresponding number of species per each sub-sample are given in Table 1. From this table, it is clear that majority of the substrate is sand. Thirty six species were identified from two core samples (https://doi.org/10.6084/m9.figshare.6541211). They belong to 32 genera, 17 families, 6 super families, 4 suborders and two orders of the class Ostracoda.

3.1 Assemblage and Vertical Distribution

The Pullivasal Island core revealed the following families to be represented by at least 3 species: *Loxoconchidae, Xestoleberididae, Leptocytheridae* and *Candonidae*. The

Table 1 Sediment parameters of two core samples (C-1) Pullivasal and (C-2) Kurusadai Islands

S.No.	Depth (cm)	$CaCO_3$ (%)		OM (%)		Substrate	
		C-1	C-2	C-1	C-2	C-1	C-2
1	2.5	88.5	90.5	1.5	1.1	Silty sand	Silty sand
2	5.0	86.5	88.5	0.6	0.9	Sand	Sand
3	7.5	85.5	86.5	0.4	0.7	Sand	Sand
4	10.0	87.0	87.0	0.5	0.6	Sand	Sand
5	12.5	88.5	79.5	0.2	0.9	Sand	Sand
6	15.0	89.0	88.0	0.5	0.9	Sand	Sand
7	17.5	78.5	82.0	0.3	0.9	Sand	Sand
8	20.0	79.5	86.5	0.4	0.9	Sand	Sand
9	22.5	85.0	93.5	0.3	0.7	Sand	Sand
10	25.0	72.5	89.0	0.3	1.0	Sand	Sand
11	27.5	90.5	84.5	0.4	0.8	Silty sand	Silty sand
12	30.0	83.5	86.5	0.2	1.0	Silty sand	Silty sand
13	32.5	79.5	84.0	0.3	0.7	Sand	Sand
14	35.0	93.5	87.0	0.6	0. 5	Sand	Sand
15	37.5	89.0	89.5	0.2	0.7	Sand	Silty sand
16	40.0	95.0	93.5	0.5	0.7	Sand	Sand
17	42.5	93.0	97.0	0.4	1.2	Sand	Sand
18	45.0	91.5	90.5	0.6	0.7	Silty sand	Sand

(continued)

Table 1 (continued)

S.No.	Depth (cm)	$CaCO_3$ (%)		OM (%)		Substrate	
		C-1	C-2	C-1	C-2	C-1	C-2
19	47.5	91.5	96.5	0.4	0.6	Sand	Sand
20	50.0	92.5	97.5	0.7	0.7	Sand	Silty sand
21	52.5	92.0	95.0	0.7	0.8	Sand	Sand
22	55.0	89.5	93.5	0.8	1.1	Sand	Sand
23	57.5	90.0	89.0	0.5	0.9	Sand	Sand
24	60.0	91.0	90.5	0.9	0.9	Sand	Sand
25	62.5	83.5	89.5	1.2	1.4	Silty sand	Sand
26	65.0	88.5	92.5	0.8	1.4	Sand	Sand
27	67.5	87.5	84.0	0.6	0.9	Sand	Sand
28	70.0	83.5	84.5	0.5	0.9	Silty sand	Sand
29	72.5	90.0	90.5	0.5	0.8	Sand	Sand
30	75.0	94.5	95.0	0.6	0.8	Sand	Silty sand
31	77.5	88.5	91.5	0.4	0.6	Sand	Sand
32	80.0	90.5	89.5	0.4	0.7	Sand	Sand
33	82.5	95.0	88.0	0.5	1.0	Silty sand	Sand
34	85.0	87.0	86.0	0.2	1.0	Silty sand	Sand
35	87.5	86.0	83.5	0.3	0.8	Sand	Silty sand
36	90.0	82.0	*	0.6	*	Silty sand	*
37	92.5	92.5	*	0.4	*	Sand	*
38	95.0	90.5	*	0.5	*	Sand	*
39	97.5	83.0	*	0.7	*	Sand	*
40	100.0	89.0	*	0.5	*	Silty sand	*
41	102.5	87.5	*	0.8	*	Sand	*
42	105.0	85.0	*	0.6	*	Sand	*
Max	105.0	95.0	97.5	1.5	1.4		
Min	2.5	72.5	79.5	0.2	0.6		
Avg	53.8	87.6	89.1	0.5	0.9		

*Samples could not be retrieved

Kurusadai Island core revealed the following families to be represented by at least 3 species: *Cytherellidae, Loxoconchidae, Bairdidae* and *Leptocytheridae*.

The vertical distribution in Pullivasal Island shows that the sub-samples from 0–22.5, 32.5–45 and 90–105 cm have 11–20 species ('high-species count samples'), whereas the rest of the samples only have 5–10 species. In Kurusadai Island the sub-samples from 0–20 cm and 40–50 cm have 11–20 species ('high-species count samples'), whereas the rest of the samples have <10 species. The higher frequency of high-species count samples at Kurusadai Island (56%) when compared to Pullivasal Island (50%) may be taken as an indication for better environmental conditions at Kurusadai Island than at Pullivasal Island. Apart from corals, mangrove plants support the congenial environment at Kurusadai Island as reported from other areas off Rameswaram (Baskar et al. 2013; Hussain et al. 2007; Maniyarsan 2016).

3.2 Cosmopolitan Species

In Pullivasal Island the species that occur in almost all the sub-samples are *Loxocorniculum lilligeborgii, Loxoconcha mandviensis, Tanella gracilis, Bairdoppilata (B.) alcyonicola, Aurila* sp., *Stigmatocythere kingmai, Xestoleberis variegata, Mutilus pentoekenesis, Loxoconcha megaporaindica* and *Lankacythere coralloides*. Species occurring only at Kurusadai Island in almost all sub-samples are *Propontocypris (S.) bengalensis, Stigmatocythere indica,*

Paijenborchellina prona, Neosionocythere dekrooni and Cytherella semitalis. These species are considered to be cosmopolitan species as they are thriving in all different environments that existed during different times especially, during the time of deposition.

3.3 Abundant Species

In Pullivasal Island, *Loxocorniculum lilligeborgii, Xestoleberis variegata, Aurila* sp., *Tanella gracilis, Mutilus pentoekenesis, Loxoconcha mandviensis* and *Loxoconcha megaporaindica* are found to occur in almost all the samples and the number of specimens for each species is more than 15. As their population counts more compared to other species and as they occur in almost all the sub-samples, they are considered abundant. *Phlyctenophora orientalis, Chrysocythere keiji, Paijenborchellina prona* and *Neonesidea dekrooni* are considered as common as their population is between 5 and 15 for each species in most of the samples. *Neonesidea cracenticlavula, Cytheropetron* sp., *Spinoceratina spinusa* and *Keijella nealei* are considered as rare species as their population is <5 for each species.

In Kurusadai Island *Loxocorniculum lilligeborgii, Xestoleberis variegata, Loxoconcha mandviensis, Tanella gracilis* and *Lankacythere coralloides* are abundant. The species *Bairdoppilata (B.) alcyonicola, Cytherella leroyi, Aurila* sp., *Mutilus pentoekenesis* and *Stigmatocythere kingmai* are common. And the species *Hemicytheridea reticulata, Macrocyprina decora* and *Paracytheroma ventrosinuosa* are rare (Maniyarsan 2016).

3.4 Species Diversity

In Pullivasal Island, the families *Trachyleberididae* and *Cytherellidae* each has a representation of 7 species and it is inferred that these families have the highest species diversity in almost all the analyzed sub-samples. *Stigmatocythere indica, Stigmatocythere kingmai, Lankacythere coralloides, Mutilus pentoekensis, Keijella reticulata, Chrysocythere keiji,* and *Keijella nealei* of the family *Trachyleberididae* are present in almost every sub-sample. *Cytherella dictyon, C. semitalis, Cytherelloidea leroyi, Hemicytheridea paiki, H. reticulata, H. khoslai* and *Keijcyoidea praecipua* of the family Cytherellidae were identified. The sub-samples of core intervals 10–15, 50.0–57.5 and 87.5–105 cm recorded the highest numbers of species from these two families, and it is inferred that the sediments deposited in this column have highest species diversity and the sediment parameters and type of substrate are more congenial for these species to thrive. Similar results have been reported from other areas off Rameswaram (Sridhar et al. 2011).

In Kurusadai Island, Families Trachyleberididae and Cytherellidae, each has a representation of 7 species, as reported in Pullivasal Island, but the number of the sub-samples Is greater than that reported in Pullivasal Island. The highest species diversity is observed between 0 and 20 cm, and between 40 and 60 cm, as well.

In Pullivasal Island, the sub-samples that recorded 11–20 species have the following environmental parameters: $CaCO_3$ content ranges from 78.5 to 95%; the organic matter is from 0.5 to 1.2% and the substrate is sand. In Kurusadai Island, the sub-samples that recorded 11–20 species have the following environmental parameters: $CaCO_3$ ranges from 79.5 to 97.5%; Organic matter is from 0.6 to 1.1% and the substrate is sand. Considering the vertical distribution of Ostracoda in terms of number of species and specimens in both islands, the $CaCO_3$ ranges from 78.5 to 97.5%; Organic matter from 0.5 to 1.2% and sand being the only substrate. These environmental parameters might have facilitated the higher numbers of species in the study area. Regarding $CaCO_3$, the higher percentages are due to coral sand. The species diversity in terms of higher number of species for the sub-samples at Pullivasal Island shows $CaCO_3$ from 83 to 92.5%; Organic matter from 0.2 to 0.8%; and the substrate is sand.

The species diversity at Kurusadai Island shows the following parameters: $CaCO_3$ from 82 to 97.5%; Organic matter 0.6–1.2%; and substrate Sand. Considering both islands, the following parameters are more accommodative for species diversity in the study area: $CaCO_3$ from 82 to 97.5%; Organic matter 0.2–1.2%; and the substrate is sand.

4 Discussion

Loxocorniculum lilligeborgii, Xestoleberis variegata, Tanella gracilis, and *Loxoconcha mandviensis* are found to be abundant at both islands. It is inferred that these species are more sustainable species to thrive in this environment (Whatley and Quanhong 1988). In Pullivasal Island, the sub-samples of core intervals from 10–15, 50.0–57.5 and 87.5–105 cm recorded the highest numbers of species and it is inferred that the sediments deposited in this column have the highest species diversity and the sediment parameters and type of substrate are more congenial for these species to thrive. Similar results have been reported from other areas off Rameswaram (Sridhar et al. 2011).

In Kurusadai Island, Families Trachyleberididae and Cytherellidae, each has a representation of 7 species, as reported in Pullivasal Island, the highest species diversity is observed between 0 and 20 cm; as well as between 40 and 60 cm.

From the species diversity, it is inferred that the Kurusadai Island has a more suitable environment than the Pullivasal Island in terms of species due to the fact that both islands have corals (Baskar et al. 2013), but Kurusadai has mangroves together with corals.

5 Conclusions

Based on the vertical distribution of recent benthic Ostracoda, it is concluded that the environmental conditions are better conserved in Kurusadai than in Pullivasal Island in terms of totally banned bottom net trawling. Apart from corals, mangrove plants support the congenial environment at Kurusadai Island as it shows higher numbers of species as well as abundance. Among the 37 species identified, thirteen species are found to be cosmopolitan. *Loxocorniculum lilligeborgii, Xestoleberis variegata, Tanella gracilis,* and *Loxoconcha mandviensis* are found to occur as abundant in both islands. It is inferred that these species are more sustainable to thrive in this environment. From the species diversity, it is inferred that the Kurusadai Island has a more suitable environment than the Pullivasal Island due to the fact that both islands have corals, but Kurusadai has mangrove plants as well as corals.

References

Baskar, K., Sridhar, S.G.D., Hussain, S.M., Solai, A., Kalaivanan, R.: Distribution of recent benthic Ostracoda off Rameswaram, Palk Strait, Tamilnadu, Southeast Coast of India. J. Geol. Soc. India (Spec. Publ.) 1, 195–212 (2013)

Gaudette, H.E., Flight, W.R., Toner, L., Folger, D.W.: An inexpensive titration method for determination of organic carbon in recent sediments. J. Sed. Res. **44**(1), 249–253 (1974)

GOMMBRE.: Gulf of Mannar Marine Biosphere Reserve Report, 46p. Tamilnadu Forest Department, Ramnathpuram (1997)

Government of India, Department of Ocean Development.: Resources Information System for Gulf of Mannar (India), 294p. Published by Integrated Coastal and Marine area Management Project Directorate, Chennai (2001)

Hussain, S.M., Ganesan, P., Ravi, G., Mohan, S.P., Sridhar, S.G.D.: Distribution of Ostracoda in marine and marginal marine habitats of Tamilnadu and adjoining areas, Southern east Coast of India and Andaman Island: environmental implications. Ind. J. Mar. Sci. **36** (4), 369–377 (2007)

Kathiresan, K.: Conservation and management of mangroves in South Asia International Conference and Exhibition on Mangroves of Indian and Western Pacific Oceans. ICEMAN, pp. 61–84 (2006)

Krumbein, W.C., Pettijohn, F.J.: Manual of Sedimentary Petrography, 549p. D. Appleton Century Co. Inc., New York (1938)

Maniyarsan, S.: Systematics, distribution and ecology of recent benthic Ostracoda, off Kurusadai Island, Gulf of Mannar, Southesast Coast of India. Ph.D. thesis, University of Madras (2016)

Piper, C.S.: Soil and Plant Analysis. University of Adelaide Press, Aldelaide. 368p, (1947)

Sridhar, S.G.D., Baskar, K., Hussain, S.M., Solai, A. Kalaivanan, R.: Distribution and diversity of recent benthic Ostracoda off Rameswaram Island, Southeast Coast of India, Tamilnadu. In: Proceedings of 17th Conversation, Indian Geological Congress, pp. 283–298 (2011)

Trefethen, J.M.: Classification of sediments. Am. J. Sci. **248**(1), 55–62 (1950)

Whatley, R., Quanhong, Z.: Recent Ostracoda of the Malacca straits part I. Rev. Exp. de Micropa **20**(1), 5–37 (1988)

Ecological Interactions of Miocene to Pleistocene Siwalik Bovids, Suids and Giraffids Traced Out by Enamel Hypoplasia Analysis

Abdul Majid Khan, Rana Manzoor Ahmad, and Ayesha Iqbal

Abstract

Enamel hypoplasia is the thinning of enamel caused by depletion of ameloblasts (enamel forming cells) during tooth development. The depletion of ameloblasts is associated with the physiological or environmental stress that an animal faced during its life history, so this dental defect can provide us a permanent and reliable record of the ecological stress faced by an animal during its tooth development. The fossils discovered from the Siwalik deposits of Pakistan having chronological age from 18.3 to 0.6 Ma and belonging to thirteen bovid, eight suid and seven giraffid species of the Siwalik regions were analyzed for comparative occurrence of enamel hypoplasia. The moderate level of stress throughout the Siwaliks for all the three studied families was inferred by enamel hypoplasia results. This shows that even though the studied three Siwalik artiodactyle families have diversity in their ecological behavior, the changing climatic condition throughout the Siwaliks had strong impact on the species of all the three studied taxa.

Keywords

Stress • Ameloblast • Dental defect • Climatic changes Artiodactyls

1 Introduction

Enamel defect studies can provide a unique perspective of environmental stress period present during the growing days of an extinct animal's life. Enamel is the hardest tissue of body and marks of Enamel Hypoplasia (EH) remain unaltered even during fossilization so it can be a highly admirable stress marker for the life history of an extinct animal.

EH is a type of enamel defect characterized by the thinning of the enamel (Goodman and Rose 1990). The formation of enamel can be categorized into two phases; the first is the secretary phase and the second is the maturation phase (Hillson 1986). In the secretary phase enamel first caps top of the tooth crown and then move down along the sides of the crown. In the maturation phase the process of mineralization occurs on the tooth (Fig. 1). EH occurs during the secretary phase.

Areas missing enamel, single or multiple pits, vertical or horizontal groves are the three types of EH described by the FDI (Federation Dentaire International 1982). The area missing enamel has advantage over the other two types in that it can be examined macroscopically. The age of an animal during a particular stress episode in its ecosystem can be accessed on the basis of the position of the area missing enamel on tooth crown relative to the tooth-crown junction (Goodman et al. 1980; Suckling 1989). There are two types of area missing enamel, Linear Enamel Hypoplasia (LEH) and Semicircular Enamel Hypoplasia (SEH). LEH is typically visible on a tooth's surface as one or more horizontal grooves and SEH is a tooth depression in semicircular form in horizontal direction (Goodman and Rose 1991; Skinner and Goodman 1992). EH is caused by a physical disruption in the ameloblasts during the tooth development. This defect is usually correlated by a systemic stress (Franz-Odendaal et al. 2004). These systemic stresses mainly caused by the deficiency of food/nutrients. The feeding deficiency is directly linked to the physiological or environmental stress. EH was analyzed in the Siwalik bovids, suids and giraffids of Pakistan in this study.

2 Materials and Methods

The analyzed materials include family Bovidae (294 teeth of 152 individuals belong to 12 genera and 13 species), family Suidae (179 teeth of 112 individuals belong to 06 genera and

A. M. Khan (✉) · R. M. Ahmad · A. Iqbal
Department of Zoology, University of the Punjab, Lahore, Pakistan
e-mail: majid.zool@pu.edu.pk

H. Chenchouni et al. (eds.), *Exploring the Nexus of Geoecology, Geography, Geoarcheology and Geotourism: Advances and Applications for Sustainable Development in Environmental Sciences and Agroforestry Research*, Advances in Science, Technology & Innovation, https://doi.org/10.1007/978-3-030-01683-8_18

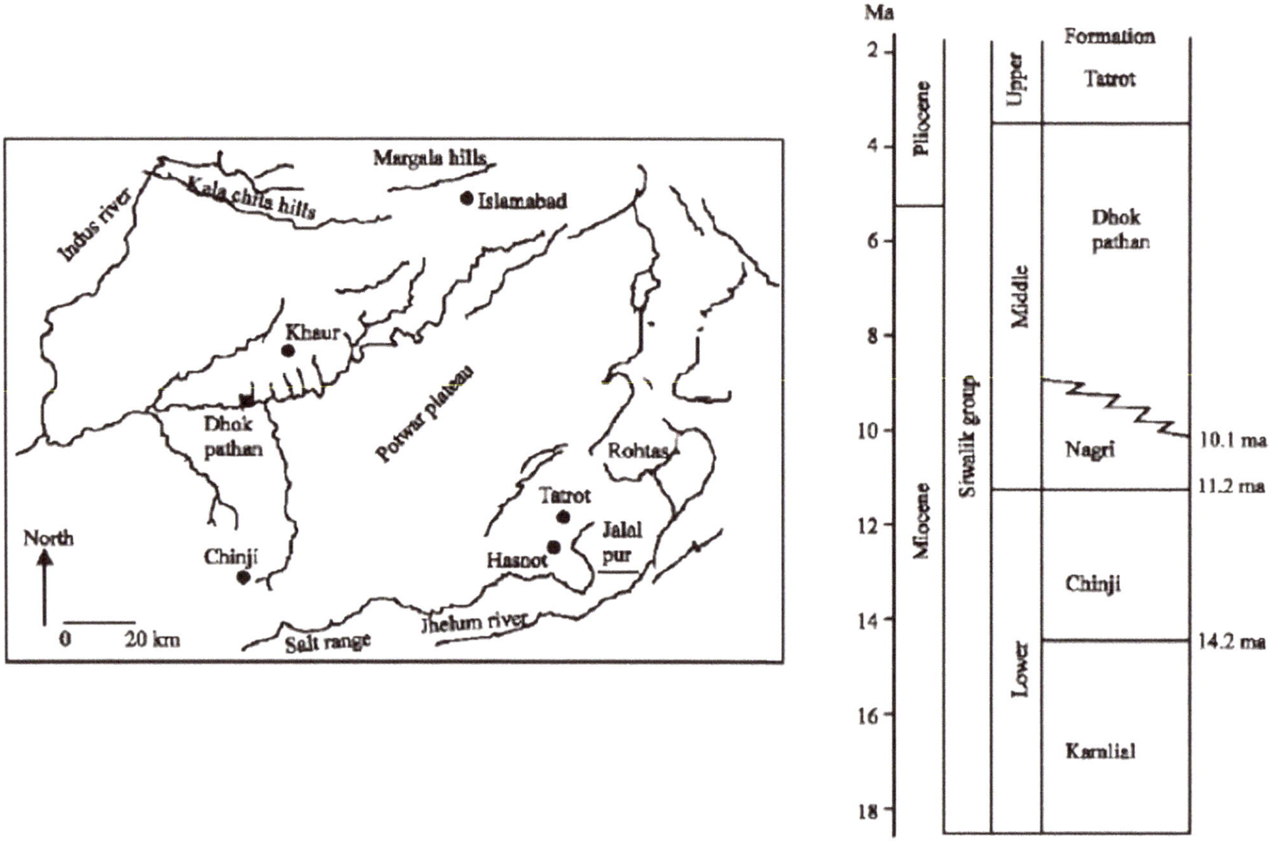

Fig. 1 Siwalik of Pakistan stratigraphic sections. Boundary dates are according to Barry et al. (2002)

08 species) and family Giraffidae (95 teeth of 58 individuals belong to 05 genera and 07 species). The samples were collected from the Siwalik out croups of Pakistan (Fig. 1). These Siwalik outcrops have early Miocene-early Pleistocene age. Severely damaged or worn teeth were excluded and only well preserved teeth were evaluated. The fossils used for analysis of EH are housed at the Dr. Abu Bakar Fossil Display and Research Center, Department of Zoology, University of the Punjab Lahore, Pakistan. Each tooth was examined visually using magnifying glasses for the presence or absence of EH. EH analysis was based on Mead (1999). The EH occurrence results are statistically analyzed using chi square test.

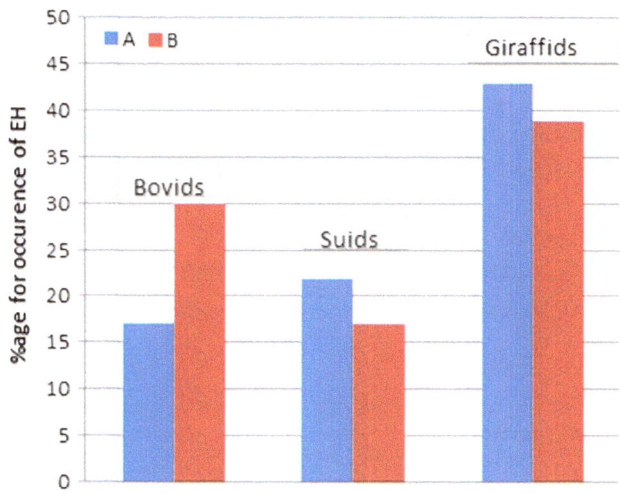

Fig. 2 Graphical representation of the comparative occurrence of enamel hypoplasia in the Siwalik bovids, suids and giraffids. For bovids A = early Miocene-early Pliocene, B = early Pliocene-early Pleistocene, for suids A = early-middle Miocene, B = late Miocene-late Pliocene and for giraffids A = early-middle Miocene, B = late Miocene-early Pleistocene

3 Results

The EH occurrence results are given in Fig. 2. This figure points out that all the three studied Siwalik artiodactyle families have a moderate level of stress throughout their existence in the Siwalik of Pakistan. For bovids, the statistical result is $\chi^2 = 3.6517$, $df = 1$, $p = 0.5601$, for suids this value is $\chi^2 = 0.4578$, $df = 1$, p = 0.4986 and for giraffids this value is $\chi^2 = 0.098$, $df = 1$, $p = 0.7545$. The difference in occurrence of EH in bovids, suids and giraffids with reference to different geological intervals of the Siwalik region of Pakistan is non-significant ($p < 0.05$).

4 Discussion

The moderate level of stress faced by the Siwalik giraffids, suids and bovids, inferred on the basis of current EH occurrence results in these animals, might be due to following stress events of the Siwalik region. Out of these three families that suffered with moderate level of environmental stress suids have competitively less occurrence of EH that might be due to the reason that the diversity in dietary habitat and omnivorous nature of suids helped them to some extend to overcome the vegetational and environment stress events in the Siwalik region.

The early Miocene was the time for transition of Paleogene to Neogene deposits, one of the major climatic turnovers. There were rapid and irregular climate changes during 17 Ma. The tectonic factors of early Miocene origin provided a noteworthy threshold for the Asian climate reorganization. The middle Miocene is a period of stress due to rapid and irregular climate changes at 13 Ma resemble that of 17 Ma changes. The increased ratio of disappearance and appearance during this period indicates a rapid change in the environment that ultimately was responsible for nutritional and ecological stress to the middle Miocene species (Barry and Flynn 1990; Barry et al. 2002).

Average species duration decreased significantly in the late Miocene due to the increase frequency of environmental disturbances. There was a very marked cooling in late Miocene after 6.5 Ma, this cooling episode had a vital role in the development of these environmental changes (Barry and Flynn 1990). There was an increase in cooling trends and glacial cycle amplitude in Plio-Pleistocene time span. The asymmetric saw tooth type glaciation pattern of these ages was one of the motives for climate dynamics. 2.5 Ma transition is considered as an important contributor to this climate dynamics (Lisiecki and Raymo 2007). According to Dennell et al. (2006) there were shifts towards more arid conditions during 1.8, 1.7 and 1.0 Ma in the Siwalik region.

5 Conclusion

EH presence in fossils of studied bovid, suid and giraffid species of Neogene deposits of the Siwaliks indicates that the number of vegetation and climatic changes in these ecosystems had strong impact on the life of the animals belonging to these artiodactyle families throughout their existence in the Siwalik region of Pakistan that had a major contribution to the evolution, extinction and migration of extinct Himalayan species. Other regions of the Siwalik may show the same trend of EH occurrence in these artiodactyles due to inter region resemblance of climatic conditions during the Miocene-Pleistocene epochs. The current results for enamel hypoplasia analysis depict that the Siwalik bovid, suids and giraffids faced moderate level of ecological stress throughout their existence in the Siwalik region of Pakistan.

References

Barry, J.C., Flynn, L.J.: Key biostratigraphic events in the Siwalik sequence. In: European Neogene mammal chronology, pp. 557–571. Springer US. (1990)

Barry, J.C., Morgan, M.E., Flynn, L.J., Pilbeam, D., Behrensmeyer, A. K., Raza, S.M., Khan, I.A., Badgley, C., Hicks, J., Kelley, J.: Faunal and environmental change in the late Miocene Siwaliks of Northern Pakistan. Paleobiol. Memoir. **28**(3), 1–72 (2002)

Dennell, R.W., Coard, R., Turner, A.: The biostratigraphy and magnetic polarity zonation of the Pabbi Hills, Northern Pakistan: an upper Siwalik (Pinjor stage) upper Pliocene-Lower Pleistocene fluvial sequence. Palaeogeogr. Palaeoclimatol. Palaeoecol. **234**(2), 168–185 (2006)

Federation Dentaire International: An epidemiological index of developmental defects of dental enamel (DDE Index). Int. Dent. J. **32**, 159–167 (1982)

Franz-Odendaal, T.A., Chinsamy, A.., Lee-Thorp, J.: High prevalence of enamel hypoplasia in an early Pliocene giraffid (*Sivatherium hendeyi*) from South Africa. J. Vert. Paleontol. **24**(1), 235–244 (2004)

Goodman, A.H., Armelagos, G.J., Rose, J.C.: Enamel hypoplasias as indicators of stress in three prehistoric populations from Illinois. Hum. Biol. **52**, 515–528 (1980)

Goodman, A.H., Rose, J.C.: Assessment of systemic physiological perturbations from dental enamel hypoplasias and associated histological structures. Yearb. Phys. Anthropol. **33**(11 Suppl), 59–110 (1990)

Goodman, A.H., Rose, J.C.: Dental enamel hypoplasias as indicators of nutritional status. In: Kelly, M. Larsen, C. (eds.) Advances in Dental Anthropology, pp. 279-293. Wiley-Liss, New York (1991)

Hillson, S.: Teeth 1st ed. Cambridge University Press, Cambridge (1986)

Lisiecki, L.E., Raymo, M.E.: Plio-Pleistocene climate evolution: trends and transitions in glacial cycle dynamics. Quat. Sci. Rev. **26**(1), 56–69 (2007)

Mead, A. J.: Enamel hypoplasia in Miocene rhinoceroses (*Teleoceras*) from Nebraska: evidence of severe physiological stress. J. Vert. Paleontol. 19(2), 91–397 (1999)

Skinner, M., Goodman, A.H.: Anthropological uses of developmental defects of enamel. In: Saunders, S.R., Katzenberg, M.A. (eds.) Skeletal biology of past peoples: research methods, pp. 153–174. Wiley-Liss, Inc., New York (1992)

Suckling, G.W.: Developmental defects of enamel – historical and present-day perspectives of their pathogenesis. Adv. Dent. Res. **3** (2), 87–94 (1989)

Why Anthracotheroides had Faced Family Level Extinction: Enamel Hypoplasia an Answer to the Question

Rana Manzoor Ahmad, Abdul Majid Khan, Amtur Rafeh, Ayesha Iqbal, and Ghazala Roohi

Abstract

The extinction of the Siwalik anthracotheroides is a family level extinction. In order to point out the possible causes of this extinction the occurrence of enamel hypoplasia is compared in 848 teeth of 487 extinct animals belonging to 39 species of 7 Siwalik artiodactyle families. The stratigraphic range of the analyzed material is from early Miocene-Pleistocene. The results show that the occurrence of enamel hypoplasia is in highest percentage in anthracotheroides among the Siwalik artiodactyle families and occurrence of enamel hypoplasia is also high in the anthracotheroides throughout their stratigraphic range in the Siwalik region. The high occurrence of enamel hypoplasia means high level of ecological stress faced by an animal so the current enamel hypoplasia analysis results indicate that the anthracotheroides had faced the highest ecological stress among all the Siwalik artiodactyls. The cumulative effect to these unfavorable ecological conditions might be the cause to this family level extinction.

Keywords

Stress • Ameloblast • Dental defect • Climatic changes Artiodactyls

1 Introduction

Enamel defect studies can provide a unique perspective of environmental stress period present during the growing days of an extinct animal's life. Enamel is the hardest tissue of body and marks of Enamel Hypoplasia (EH) remain unaltered even during fossilization so it can be a highly admirable stress marker for the life history of an extinct animal.

EH is a type of dental defect characterized by the thinning of enamel (Goodman and Rose 1990). The formation of enamel can be categorized into two phases, first is known as secretary phase while the second is called the maturation phase (Hillson 1986). In the secretary phase enamel first caps top of the tooth crown and then move down along the sides of the crown. In the maturation phase the process of mineralization occurs on the tooth (Fig. 1). EH occurs during the secretary phase.

Areas missing enamel, single or multiple pits, vertical or horizontal groves are the three types of EH described by the FDI (Federation Dentaire International 1982). The area missing enamel has advantage over the other two types in that it can be examined macroscopically. The age of an animal during a particular stress episode in its ecosystem can be accessed on the basis of position of area missing enamel on tooth crown relative to the tooth-crown junction (Goodman et al. 1980; Suckling 1989). There are two types of area missing enamel, Linear Enamel Hypoplasia (LEH) and Semicircular Enamel Hypoplasia (SEH). LEH is typically visible on a tooth surface as one or more horizontal grooves and SEH is a tooth depression in semicircular form in a horizontal direction (Goodman and Rose 1991; Skinner and Goodman 1992). EH is caused by a physical disruption in the ameloblasts during the tooth development. This defect is usually correlated by a systemic stress (Franz-Odendaal et al. 2004). These systemic stresses are mainly caused by the deficiency of food/nutrients. The feeding deficiency is directly linked to the physiological or environmental stress. EH is analyzed in the Siwalik anthracotheroides of Pakistan in this study.

R. M. Ahmad · A. M. Khan (✉) · A. Rafeh · A. Iqbal
Department of Zoology, University of the Punjab, Lahore, Pakistan
e-mail: majid.zool@pu.edu.pk

G. Roohi
Pakistan Museum of Natural History (PMNH), Islamabad, Pakistan

H. Chenchouni et al. (eds.), *Exploring the Nexus of Geoecology, Geography, Geoarcheology and Geotourism: Advances and Applications for Sustainable Development in Environmental Sciences and Agroforestry Research*, Advances in Science, Technology & Innovation, https://doi.org/10.1007/978-3-030-01683-8_19

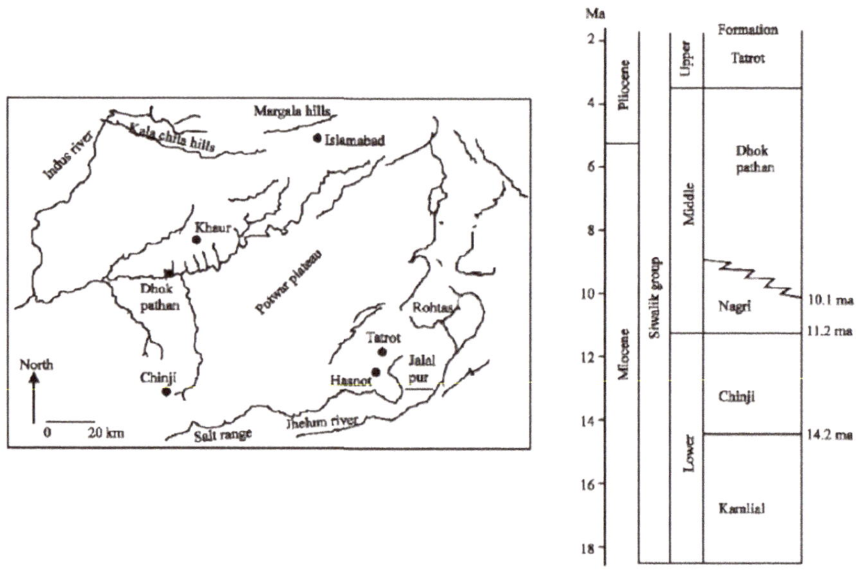

Fig. 1 Siwalik of Pakistan stratigraphic sections. Boundary dates are according to Barry et al. (2002)

2 Materials and Methods

The studied materials included 848 teeth of 487 extinct animals belong to 39 species of 07 Siwalik artiodactyle families. The samples were collected from the Siwalik out croups of Pakistan (Fig. 1). These Siwalik outcrops have early Miocene-early Pleistocene age. Severely damaged or worn teeth were excluded and only well preserved ones were evaluated. Each tooth was examined visually using magnifying glasses for the presence or absence of EH. EH analysis was based on Mead (1999). The EH occurrence results were statistically analyzed using chi square test.

3 Results

The EH occurrence results of the studied material of the Siwalik artiodactyle families of Pakistan (Fig. 2) show that the percentage for occurrence of EH is highest in the animals of family Anthracotheriidae among these families. The results for occurrence of EH in the anthracotheroides of two different stratigraphic ranges are graphically represented in Fig. 3. The results of EH in the Siwalik artiodactyls depicts that the anthracotheroides had no significant difference in the occurrence of EH throughout their range in the Siwaliks according to the statistical analysis of the current results ($p > 0.05$) and has highest frequency of EH among the Siwalik artiodactyls. This high EH is due to the following unfavorable ecological stresses faced by the anthracotheroides.

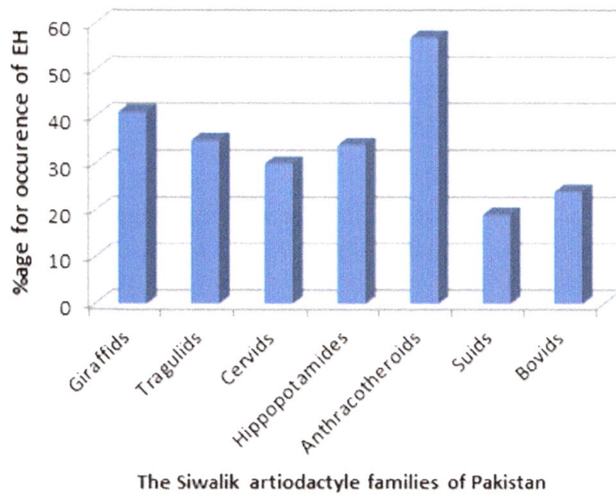

Fig. 2 Graphical representation of the family level comparison for occurrence of enamel hypoplasia in the Siwalik artiodactyls of Pakistan

4 Discussion

The depletion of the water bodies was the major stress precursor for the Siwalik anthracotheroides. According to Khan et al. (2014) humidity was high in the middle Miocene Siwalik environment, but the water availability decreased in the late

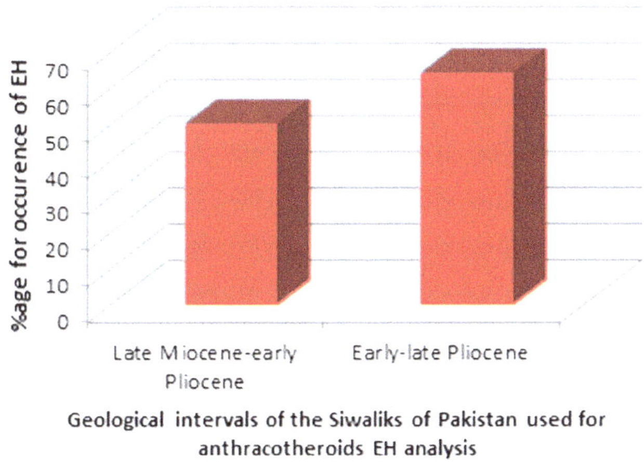

Fig. 3 Graphical representation of the comparative occurrence of enamel hypoplasia in the Siwalik anthracotheroides of Late Miocene-early Pliocene and early-late Pliocene epochs

Miocene to onward Siwalik habitat due to decreased precipitation. Quade et al. (1989) also reported dry environmental condition after 9 Ma. The dependence of anthracotheroides on water resources can be explained on the basis of their affinities with aquatic life style bearing mammals as Colbert (1935) reported strong resemblance of anthracotheroides with members of family Hippopotamidae. Some members of the anthracotheroides family were also terrestrial according to Lihoreau (2003) but they were very early representative of the family and also very much small in their size. In spite of this terrestrial nature of some species limited availability of water was significant stressor for anthracotheroides genera specially *Merycopotamus* as mentioned by Gratiollet (1867). Lihoreau et al. (2007) narrated that elevation of the supraorbital process and auditory organs in late Miocene to onward species of genus *Merycopotamus* are the morphological evidence for aquatic adaptation in this genus.

Along with limited water availability the other significant factor that played a role in the extinction of anthracotheroides was predatory stress. The imbalanced increase in weight as compared to height with less cursorial limbs and more massive body structure developed during Miocene (Lihoreau et al. 2007) made this taxon more viable for predators. The other reason for anthracotheroides extinction was the increased interspecific competition among anthracotheroides due to narrow geographical range of the family as both Black (1972) and Gaziry (1987) reported the existence of *Merycopotamus* (most common genus of family Anthracotheriidae) in Asian region only.

5 Conclusion

For Anthracotheroides to survive under the above mentioned stressful events migration could be a good alternative in order to get rid of the unfavorable ecological biotic as well as abiotic factors. But it seems that this opportunity could not extend any favor at all to anthracotheroides as their amphibious life style made their migration highly dependent on availability of interconnected water bodies across adjacent bio-geographical regions that further apprehended their migration potential and led to global level extinction of family Anthracotheroides during Pliocene epoch. The current enamel hypoplasia results point out that the anthracotheroides faced the highest ecological stress among all the reported artiodactyle families of the Siwalik region that might have led to their family level extinction during Pliocene.

References

Barry, J.C., Morgan, M.E., Flynn, L.J., Pilbeam, D., Behrensmeyer, A.K., Raza, S.M., Khan, I.A., Badgley, C., Hicks, J., Kelley, J.: Faunal and environmental change in the late Miocene Siwaliks of northern Pakistan. Paleobiol. Mem. **28**(3), 1–72 (2002)

Black, C.C.: A new species of Merycopotamus (Artiodactyla: Anthracotheriidae) from the late Miocene of Tunisia. Notes du Service de Géologie de Tunisie **37**, 5–39 (1972)

Colbert, E.H.: Siwalik mammals in the American Museum of natural history. Trans. Am. Philos. Soc. **26**, 1–401 (1935)

Federation Dentaire International.: An epidemiological index of developmental defects of dental enamel (DDE Index). Int. Dent. J. **32**, 159–167 (1982)

Franz-Odendaal, T.A., Chinsamy, A., Lee-Thorp, J.: High prevalence of enamel hypoplasia in an early Pliocene giraffid (Sivatherium hendeyi) from South Africa. J. Vert. Paleontol. **24**(1), 235–244 (2004)

Gaziry, A.W.: *Merycopotamus petrocchii* (Artiodactyla, Mammalia) from Sahabi, Libya. In: Boaz, N., Arnauti, E.L., Gaziry, A.W., de Heinzelin, J., Boaz, D. (eds.) Neogene paleontology and geology of Sahabi, pp. 287–302. New York, NY, Allan Liss (1987)

Goodman, A.H., Rose, J.C.: Assessment of systemic physiological perturbations from dental enamel hypoplasias and associated histological structures. Year b Phys. Anthropol. **33**(11 Suppl), 59–110 (1990)

Goodman, A.H., Rose, J.C.: Dental enamel hypoplasias as indicators of nutritional status. In: Kelly, M.,Larsen, C. (eds.) Advances in Dental Anthropology, pp. 279–293, New York, Wiley-Liss (1991)

Goodman, A.H., Armelagos, G.J., Rose, J.C.: Enamel hypoplasias as indicators of stress in three prehistoric populations from Illinois. Hum. Biol. **52**, 515–528 (1980)

Gratiollet, L.P.: Recherches sur l'anatomie de l'hippopotame, p. 405. Masson et fils, Paris (1867)

Hillson, S.: Teeth 1st ed. Cambridge University Press, Cambridge (1986)

Khan, A.M., Cerdeno, E., Akhtar, M., Khan, M.A., Iqbal, A., Mubashir, M.: New fossils of *Gaindatherium* (Rhinocerotidae, Mammalia) from the Middle Miocene of Pakistan. Turkish J. Earth. Sci. **23**(4), 452–461 (2014)

Lihoreau, F., Barry, J., Blondel, C., Chaimanee, Y., Jaeger, J.J., Brunet, M.: Anatomical revision of the genus *Merycopotamus* (Artiodactyla; Anthracotheriidae): its significance for Late Miocene mammal dispersal in Asia. Palaeontology **50**(2), 503–524 (2007)

Lihoreau, F.: Systématique et paléoécologie des Anthracotheriidae [Artiodactyla; Suiformes] du Mio-Pliocène de l'Ancien Monde: implications paléobiogéographiques, p. 396. Ph.D. thesis, Université de Poitiers (2003)

Mead, A.J.: Enamel hypoplasia in Miocene rhinoceroses (Teleoceras) from Nebraska: evidence of severe physiological stress. J. Vert. Paleontol. **19**(2), 391–397 (1999)

Quade, J., Cerling, T.E., Bowman, J.R.: Development of Asian monsoon revealed by marked ecological shift during the latest Miocene in northern Pakistan. Nature **342**(6246), 163–166 (1989)

Skinner, M., Goodman, A.H.: Anthropological uses of developmental defects of enamel. In Saunders, S.R., Katzenberg, M.A. (eds.) Skeletal Biology of Past Peoples: Research Methods, pp. 153–174, Wiley-Liss, Inc., New York (1992)

Suckling, G.W.: Developmental defects of enamel—historical and present-day perspectives of their pathogenesis. Adv. Dent. Res. **3** (2), 87–94 (1989)

Part VI

Socio-economics of Geotourism and Archaeology

Urban Accessibility and Its Economic Linkage Analysis: A Case Study in Beijing, Tianjin and Hebei Region, China

Chenxi Li

Abstract
Based on the analysis of urban function intensity index, this paper applied and improved potential model to implement measurement of the urban accessibility composite and spatial differences analysis in Beijing-Tianjin-Hebei Region of China. The results showed that, the grade of urban accessibility and economic linkage intensity in Beijing-Tianjin-Hebei Region were inhomogeneous; the spatial patterns of urban accessibility and economic linkage intensity in Beijing-Tianjin-Hebei Region revealed a spatial pattern distribution of spheres structure. According to the spatial pattern distribution rule of accessibility and economic linkage intensity, this paper revealed the problem of spatial pattern structure and traffic network structure in Beijing-Tianjin-Hebei Region, explained the role that economic level, development model and location playi in the difference of spatial accessibility, and proposed policy recommendations for improving the accessibility of transport networks.

Keywords
Potential model • Economic linkage • Accessibility
Spatial pattern • Beijing-Tianjin-Hebei region

1 Introduction

The Chinese government has called for integrated and coordinated development of regions around Beijing (Rainy and Emily 2016), the capital and the second largest city of China. Tianjin, a port city with a population of 14 million, is linked to Beijing by several highways and a high-speed railway. Hebei, a traditional industrial and transportation hub surrounding both Beijing and Tianjin, is also planning to build a high-speed railway that will reduce travel time to the capital to one hour. The two north China municipalities of Beijing and Tianjin form one of China's most economically developed regions along with Hebei Province. Over 100 million people inhabit the region with an area of 216,000 km^2 (NBSC 2014). Therefore, urban development of the Beijing-Tianjin-Hebei region is more fully grown in China.

At present, according to Beijing, Tianjin and Hebei Regional construction framework, governments of Beijing city, Tianjin city and Hebei province have been promoting industrial structural adjustment and layout planning to intensify each city development (Bok and Kwon 2016; Li et al. 2016).

Based on a potential model, characteristics of spatial relationship, situations of traffic network and economic and social conditions to reveal the relationship between the spatial distribution accessibility characteristics and economic linkage characteristics in the BTH region, the present paper evaluated the comprehensive development of each city in Beijing, Tianjin and Hebei regional traffic network.

2 Materials and Methods

2.1 Study Area

The study area, which is located at 113°04–119°53′ E, 36° 01–42°37′ N, has a total area of about 217.185 km^2 and includes Beijing municipality, Tianjin Municipality and Hebei Province (BTH Region). Hebei Province includes Shijiazhuang City, Tangshan City, Langfang City, Qinhuangdao City, Chengde City, Cangzhou City, Xingtai City, Handan City, Zhangjiakou City, Baoding City and Hengshui City.

2.2 Data Collection

This paper selected all of the 13 cities located in the BTH region including Beijing city, Tianjin city and 11 cities in

C. Li (✉)
School of Management, Xi'an University of Architecture and Technology, 710055 Xi'an, China
e-mail: xywglcx@163.com

H. Chenchouni et al. (eds.), *Exploring the Nexus of Geoecology, Geography, Geoarcheology and Geotourism: Advances and Applications for Sustainable Development in Environmental Sciences and Agroforestry Research*, Advances in Science, Technology & Innovation, https://doi.org/10.1007/978-3-030-01683-8_20

Hebei province as study units. Spatial data were gathered from Open street map Database, and includes expressway, urban expressway, national highway and administrative boundary. Building BTH regional traffic network data set by using the Network Analysis module in ArcGIS 10.2, this paper gave priority to expressway and examined its New OD Cost Matrix.

2.3 Methods

The potential model can be calculated as follows:

$$A_i = \sum_{j=1}^{n} \frac{M_j}{C_{ij}^a} \tag{1}$$

where a denotes node cities' travel friction coefficient, which reflects the influence degree of a spatial and temporal barrier on the accessibility relation of any two node cities. It is usually calculated as Eq. 1.

A node city economic membership degree *Ri*, which was then calculated by using the sum of r_{ij} of each row in the matrix, reflected the economic interaction intensity with any other node city. *Ri* can be calculated as follows:

$$R_i = \sum_{j=1}^{n} r_{ij} \tag{2}$$

Furthermore, *a* node city economic linkage intensity indicator *ai*, indicated by using the ratio of the node cities average economic membership degree in the traffic network, reflected the advantages and disadvantages of the node cities location conditions in the economic network system. It can be calculated as follows:

$$a_i = \frac{R_i}{(\sum_{i=1}^{n} R_i)/n} \tag{3}$$

where the larger the value of *ai* is, the better the geographical location of the node city will be. A node city that owned the largest value of *ai* was located in the geometric center in the urban economic network system.

3 Discussion

Based on the results of node cities accessibility evaluation, this paper used Eqs. (2) and (3) to calculate the indices of urban economic ties intensity of cities in the BTH region (Table 1) and reflect the size of the economic radiation intensity of a certain node city affected by the other node cities in the study area. The results showed that *ai* of Beijing was the highest (*ai* = 4.072), therefore, Beijing was the economic geometric center in the BTH region.

Table 1 Indices of urban economic linkage intensity of cities in the BTH region

City	a_i	City	a_i	City	a_i
Beijing	4.072	Baoding	0.654	Hengshui	0.211
Tianjin	2.652	Handan	0.634	Zhangjiakou	0.207
Tangshan	1.173	Cangzhou	0.621	Qinhuangdao	0.190
Shijiazhuang	1.112	Xingtai	0.313		
Langfang	0.943	Chengde	0.217		

Based on the above analysis, using ArcGIS 10.2 Kriging spatial interpolation method, we got regional accessibility (*Ai*) spatial pattern and characteristics in the BTH region (Fig. 1a). It showed that Beijing was the geometrical center of the regional accessibility spatial distribution in the BTH region. Spatial distribution characteristics showed a gradual decreasing trend from the central area to the southeast area. In addition, *Ai* of the north areas was generally low.

Using ArcGIS 10.2 Kriging spatial interpolation method, this paper got regional urban economic linkage intensity (*ai*) spatial pattern and characteristics in the BTH region (Fig. 1b). From the regional distribution differences, similar to *Ai*, the spatial pattern and characteristics of *ai* did not show concentric circle spatial distribution structure characteristics but also a law of distance diminishing, the farther a city is away from the center of regional economy, the weaker the *ai* of the city will be. Urban economic linkage spatial differences indicated that the size of urban economic linkage intensity (*ai*) was related to the comprehensive economic radiation capacity affected by other cities and their location conditions.

4 Conclusion

This paper revealed that the grade of urban accessibility and economic linkage intensity in the BTH region ranked from high to low in order: Beijing, Tianjin, Tangshan, Shijiazhaung, Langfang, Baoding, Handan, Cangzhou, Xingtai, Chengde, Hengshui, Zhangjiakou and Qinhuangdao; the spatial patterns of urban accessibility and economic linkage intensity in the BTH region showed a spatial pattern distribution of spheres structure, i.e., the central and western areas

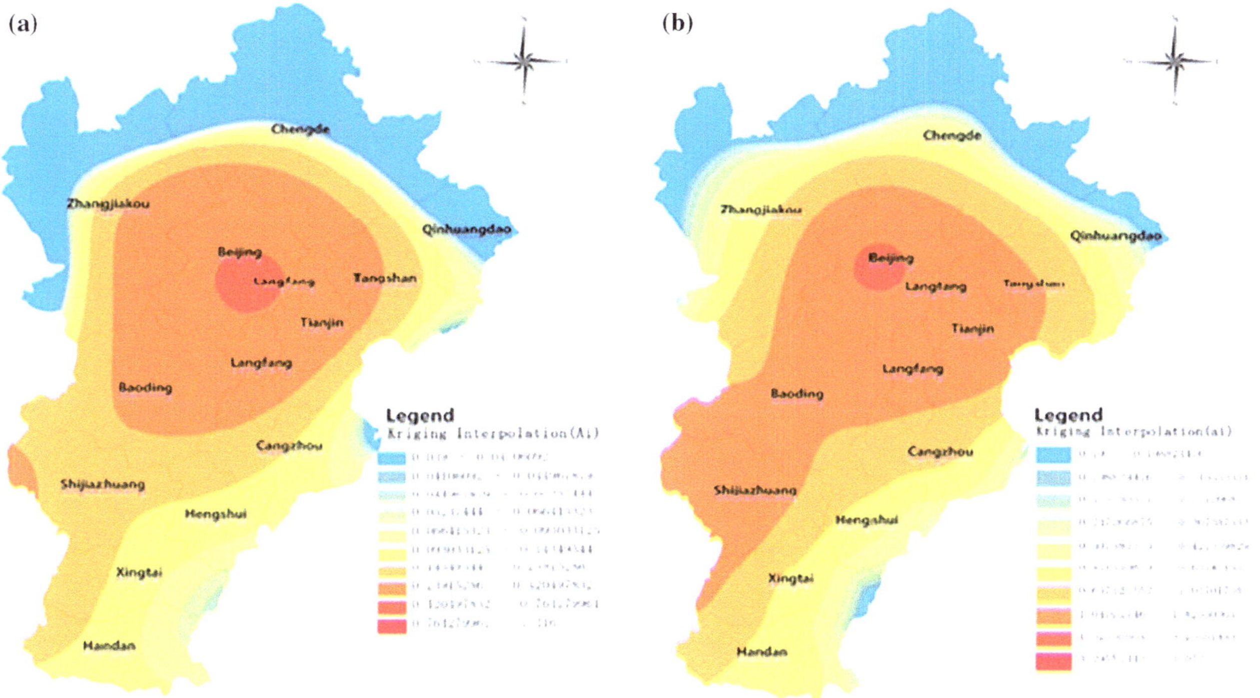

Fig. 1 **a** Regional accessibility (*Ai*) spatial pattern and characteristics in the BTH region; **b** Urban economic linkage intensity (*ai*) spatial pattern and characteristics in the BTH region

showed a relatively better accessibility and economic linkage intensity, while the northern areas showed a relatively worse situation.

We put forward the following suggestions for the future development of the BTH region. The economic linkage and cooperation among cities in the BTH region should be strengthened and the interaction between the local area and other regional economic spatial geometric centers should be promoted. In addition, the construction of an infrastructure should be strengthened, especially, improve the road network structure and constantly improve the urban road traffic system.

References

Bok, J., Kwon, Y.: Comparable measures of accessibility to public transport using the general transit feed specification. Sustainability **8** (3), 224 (2016)

Li, T., Wu, J., Sun, H., Gao, Z.: Integrated co-evolution model of land use and traffic network design. Netw. Spat. Econ. **16**, 579–603 (2016)

NBSC: China Statistics Yearbook 2014. China Statistics Press, Beijing (2014)

Rainy, Y., Emily, L.: Jing-Jin-Ji: The Biggest city in China you've probably never heard of. http://www.china-briefing.com/news/2014/07/11/jing-jin-ji-biggest-city-china-youve-probably-never-heard.html. Last accessed 3 Aug 2016

An Assessment of Socio-economic Profile and Sustainable Development in Tourism Potential Areas: A Case Study of Galiyat (Pakistan)

Razzaq Ahmed, Shaukat H. Khan, and Khalida Mahmood

Abstract

Pakistan is a country with a great potential of tourism but unfortunately this sector has remained badly neglected. The north and west of Pakistan attracts tourists from all over the world due to its eye-catching natural beauty, massive and unique glaciers, sky-scraping peaks, fast flowing rivers and historical sites related to different civilizations. In spite of these facts the socio-economic conditions of these tourist rich areas are unsatisfactory. The present study was an attempt to assess the socio-economic profile of the tourist potential region of Lesser Himalayas-The Galiyat, by a comprehensive questionnaire. Galiyat is the most attractive destination among the local tourists and hence badly affected by mass tourist and unplanned construction. This study suggested a sustainable economic development in the study area using indigenous resources and potential, and proposed suitable sites for eco-tourism development.

Keywords

Eco-tourism • Galiyat • Socio-economic profile • Sustainable development • Tourist potential

1 Introduction

Tourism plays a vital role in the socio-economic development of any area (Mock 1999). There are many countries in the world that entirely or partially depend on tourism. Pakistan has great potential and immense resources for tourism in terms of natural sceneries, historical places, desert safaris and beaches (Ahmed and Mahmood 2017). But due to the lack of advertisement, unconstructive propaganda and proper marketing at international level, many of the tourist products remain unknown and facing the spread of poverty and regional disparity (Mahmood 2004). The current study focused on exploring the tourist potential of Galiyat region which can be used for economic development and environmental sustainability.

The aims and objectives of this study were to:

- Evaluate the perception of local population about tourism in their area
- Develop a strategy for a sustainable tourism in the study area
- Identify the potential sites for an eco-tourism in Galiyat.

2 Materials and Methods

This study is an effort towards the economic sustainability and ecotourism development in the tourism rich region of Galiyat through the following methodology:

(a) Extensive field survey to gather the local perception using a comprehensive questionnaire.
(b) Propose ecotourism sites selected during the field survey.

3 Results and Discussion

Galiyat is a plural of gali which is a local terminology that means pass. There are number of gallies spread over a large area between Murree and Abbottabad and possess great natural scenic beauty (Fig. 1). Galiyat has most of the famous

R. Ahmed (✉)
Department of Geography, Federal Urdu University of Arts, Science and Technology, Gulshan Campus, Karachi, Pakistan
e-mail: razzaqahmed@fuuast.edu.pk

S. H. Khan
National Institute of Oceanography, Karachi, Pakistan

K. Mahmood
Department of Geography, University of Karachi, Karachi, Pakistan

H. Chenchouni et al. (eds.), *Exploring the Nexus of Geoecology, Geography, Geoarcheology and Geotourism: Advances and Applications for Sustainable Development in Environmental Sciences and Agroforestry Research*, Advances in Science, Technology & Innovation, https://doi.org/10.1007/978-3-030-01683-8_21

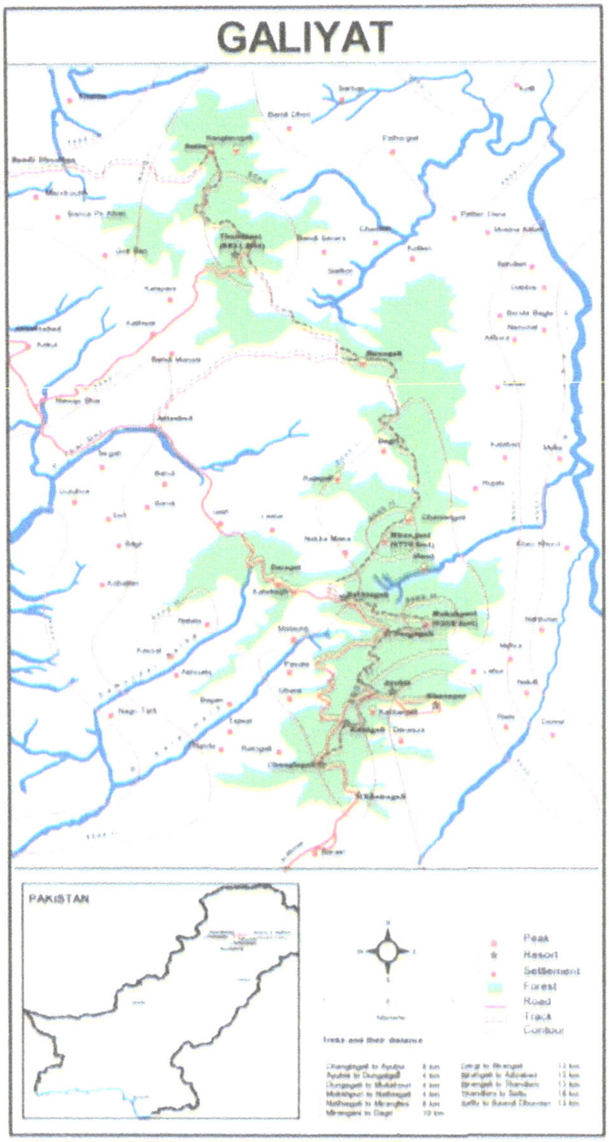

Fig. 1 Galiyat: the study area

Plate 1 Ayubia: one of the famous tourist resort in Galiyat

tourist resorts in Pakistan such a Nathiagali, Ayubia, Changlagali, Bhurban etc. (Talat and Khan 2004–2005) (Plate 1). The entire area of Galiyat is badly affected by mass tourists and needs to develop an environmental friendly tourism for the economic and environmental sustainability (Ceballos-Lascurain 1996). It has very dense forest of conifers (such as pine, chir, deodar) due to the high annual rainfall (1400 mm). The summer season is refreshingly cool and pleasant and attracts a large number of tourists not only in summer but also in winter due to the heavy snowfall. Hence, the economy of the area is mainly based on tourism. However, due to improper tourism planning and management, it is unable to get full benefits from this economic sector.

The current study is based on extensive field survey. The respondents during the survey included people covering an age group of 25–60+, involving businessmen, employees and laborers, whose education varied from below high school (36%). high school to above high school (64%). According to the survey the education ratio is high in the study area and the majority of the people are literate. Primary and secondary schools are available in almost every part of the area which includes both government and private, however higher educational institutes are not available except for summer campuses of few universities like the Punjab and Peshawar.

There is a shortage of health facilities, just a single hospital (Government General Hospital, Murree) and a CMH Murree is available; people mostly travel to Rawalpindi, Murree or Abbottabad to seek better health care. High blood pressure is very common and most of the people suffer from this disease. The other health issues include heart attack, blood sugar, stroke, and abdominal diseases.

Galiyat has very limited agricultural activities. Cash crops such as wheat, rice, sugar cane and other cereals are not cultivated for commercial purposes. Local people cultivate wheat but only for subsistence purposes and the other agricultural products include maize, fruits (such as apple, nuts, pear and peach) and vegetables (such as tomato, onion, turnip, radish, potato, lady finger, cucumbers and carrot) to fulfill their own basic needs. However, some (20–25%) of people's sales consists of their home-grown crops they sell in an open market. The results of the survey conducted regarding ownership of land, people cropping in the field, types of crops, income and other related information revealed that most of the people work on their own land. About 90.90% people work in the fields by themselves and 9.09% have laborers in their field hired to work. The average farm size is quite small, ranging between 1 and 15 acres, whereas the annual income generated from the agricultural output is between 15,000 and 200,000 PKR (124–1600 US$).

Agriculture practice in the study area is in a traditional way with very minimal use of farm machinery and chemical fertilizers. The irrigational network is not available in the study area and rain water is the only mode of irrigation which affects the agricultural products if the rain pattern is disturbed. When asked about the government compensations for the losses in agriculture due to any natural calamity, the people replied negatively.

Livestock raising and dairy activities are also closely tied to farming activities but mostly for domestic need except for 18% which is put on the market. The majority of the people use LPG for their domestic fuel, but people also use wood supplied by the government as deforestation is strongly prohibited in the study area.

Natural disasters like landslides and earthquakes are common in the study area. In the case of landslides, 82% of people answered that the government does not help, 9% said "yes, the government helps while 9% of people answered that government does sometimes help them".

Galiyat have always been the trekkers' paradise. It has some of the best treks of the world which are full of natural colors, such as Mushkpuri-Dongagali, Mushkpuri-Nahtiagali, Nathiagali-Miranjani, Dungagali-Ayubia, Birangali-Thandiani and Khanspur-Lahore Kas (Plate 2). These treks hold great potential for eco-tourism development because of its rich natural vegetation cover, forests and the entire ecosystems that not only attract numerous tourists nationally and internationally but could also be a major source of income for the indigenous communities living in the area (Rana 2002).

The natural scenery of some sites in Galiyat is being damaged by mass tourism. Murree, for instance, was converted into concrete jungle instead of a pine forest. Now the enchanted and peaceful sound of wind through the pine trees is very rare. It has been replaced by the rotting garbage, choked sewerage system, exhausted fumes and greasy smoke, if this process is not stopped, it will spread over the whole region of Galiyat (Nasir 2002).

The socio-economic analysis of the study area shows that it faces the dire consequences of underdevelopment and increasing poverty. A vast majority of people desire the most basic needs in education, health, employment and transport. The current research will help achieve the sustainable goals in Galiyat by developing economically and environmentally sound hill tracts and a mountainous ecosystem. This will certainly improve the living standards of the local residents and sustain such a vital ecosystem affecting positively those living in downstream and plain areas.

Plate 2 Khanspur-Lahore Kas trek

4 Outcomes

The unique environment and culture of Galiyat offers opportunities as well as challenges. In an era of connectivity and economic integration, this area can be developed with the integration of local community since the beginning as mentioned by Lash (1997). Their involvement in the sustainable development projects is possible through the goals of trust and confidence. To achieve the transformative change, careful attention must be given spatially and culturally especially for women. A successful implementation will necessitate fostering the stakeholders engagements among government, local communities, development agencies, private sectors, NGOs, donors, academia and research organizations.

References

Ahmed, R., Mahmood, K.: Tourism potential and constraints: an analysis of tourist spatial attributes in Pakistan. Pak. Perspect. **22**(2), 61–71 (2017)

Ceballos-Lascurain: Tourism, Eco-tourism and Protected Areas. IUCN Protected Areas Program, IUCN (1996)

Lash, G.: What is community based eco-tourism for forest conservation and community development. In: Proceeding of an International Seminar, Thailand, 28–31 January (1997)

Mahmood, K.: Dynamic of regional inequality of districts of Pakistan: 1981–2001. Pak. Perspect. **9**(2), 85–103 (2004)

Mock, J.: Eco-tourism potential in Pakistan's northern areas. Paper presented at the Convention on Sustainable Tourism in Pakistan's Northern Areas, Gilgit and Hunza, 12–14 June (1999)

Nasir, Z.: Murree committing commercial suicide. Sunday Magazine, The Dawn, 21 April (2002)

Rana, A.: Mahol dost Sayahat. Jareeda IUCN, 18–19, April–May–June (2002)

Talat, B., Khan, S.: Tourism potential and problems in mountains of Pakistan: a case study of Hazara division. Grassroots **33**, 54–60 (2004–2005)

Archaeology and Archaeometry of Ceramics of the Roman City of *Thaenae* (Tunisia): Between Inland Roman Africa and the Mediterranean Basin

Rémi Rêve, Jean-Paul Ambrosi, Claudio Capelli, Michel Bonifay, and Abdelhamid Barkaoui

Abstract

This research project is about the archaeological and archaeometric characterization of African ceramics. The development of a combined archaeological, petrographic and geochemical approach is currently one of the most promising approaches for the determination of ancient ceramic productions and their tracing throughout the Mediterranean basin. Tested on the collection of African Red Slip wares of the Museum of Gap (France), this interdisciplinary method is currently tested on the ceramic material of the Ancient city of *Thaenae*/Thyna (Tunisia). This integrated study aimed to provide a characterization of the pottery production specific to the *Thaenae* site, measure its diffusion in the Mediterranean basin and determine the economic role of *Thaenae* during late antiquity. The development of rigorous protocols would allow the application of this approach to other ceramic productions in the Mediterranean basin.

Keywords

Archaeology • Archaeometry • Roman pottery
Petrography • Portable XRF

R. Rêve (✉) · M. Bonifay
Aix Marseille University, CNRS, CCJ, 13090 Aix-en-Provence, France
e-mail: remi.reve@etu.univ-amu.fr

M. Bonifay
e-mail: michel.bonifay@univ-amu.fr

R. Rêve · A. Barkaoui
Sfax University, LERIC, 3029 Sfax, Tunisia
e-mail: hamidbarkaoui5@gmail.com

J.-P. Ambrosi
Aix Marseille University, CNRS, IRD, CDF, INRA, CEREGE BP 80, 13545 Aix-en-Provence, France
e-mail: ambrosi@cerege.fr

C. Capelli
Dipartimento di Scienze della Terra, dell'Ambiente e della Vita (DISTAV), Università degli Studi di Genova, 16126 Genoa, Italy
e-mail: capelli@dipteris.unige.it

1 Introduction

Since the second half of the 20th century and particularly in the 1990s, the archaeological characterization of African ceramics has known a strong development. It was at the beginning of 2000s that we have seen the development of petrographic and geochemical characterization (Bonifay et al. 2012) and at the end of the 2000s, an integrated approach with archaeology and petrography was introduced providing significant data for the identification of workshops and typology (Capelli et al. 2007). In the continuity of this research type, a new method was presented here, focusing on a combined approach currently used on the archaeological site of *Thaenae*/Thyna (Tunisia) and namely on the characterization of the local ceramic production. This research included a PhD thesis (R.R) which started in 2017 under the supervision of Aix Marseille University (France) and Sfax University (Tunisia).

2 Materials and Methods

This study concerns all ceramic materials produced in *Thaenae*'s workshops (12 km south of Sfax) (Fig. 1), attested after the discovery of some kiln wasters especially of amphorae and common wares during preliminary field surveys. But, it also concerns imported ceramics (especially tablewares). Therefore, a combined approach was developed to characterize these materials with a ceramic study based on mineralogical-petrographic and geochemical methods (thin-section analysis under a polarizing microscope; analysis with portable X-Ray Fluorescence spectrometer 'XRF').

This last method, very promising in archaeological issues (Frahm 2018) but still to be tested extensively (Hunt and Speakman 2015), offers many advantages related to the mobility of the equipment and its non-destructive aspect. This integrated approach was carried out for a preliminary

H. Chenchouni et al. (eds.), *Exploring the Nexus of Geoecology, Geography, Geoarcheology and Geotourism: Advances and Applications for Sustainable Development in Environmental Sciences and Agroforestry Research*, Advances in Science, Technology & Innovation, https://doi.org/10.1007/978-3-030-01683-8_22

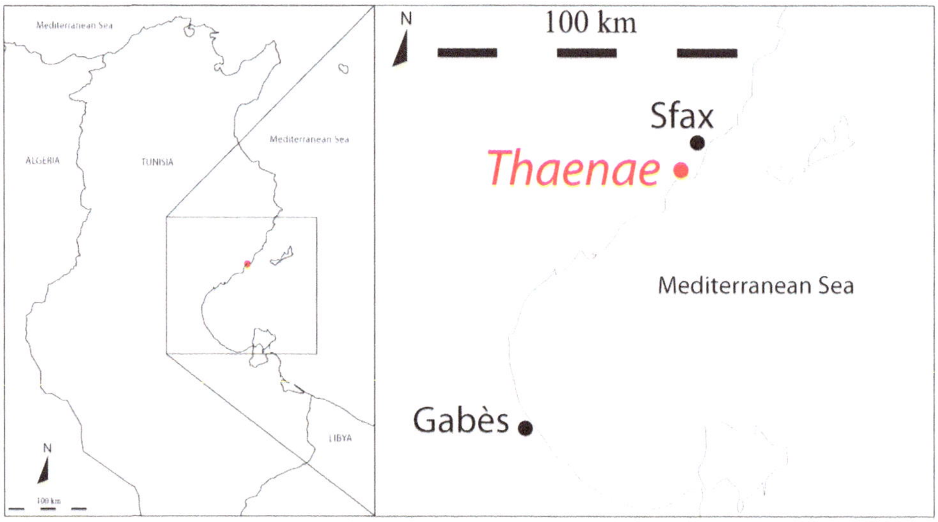

Fig. 1 Location of the archaeological site of *Thaenae*

analysis of the African Red Slip collection of the Musée Muséum Départemental de Gap (France) with rather good results (Mukaï et al. 2016). However, it is still necessary to develop cross measurements with petrography and improve the number of the analysed samples in order to validate the results and define rigorous protocols.

3 Results

We synthetically presented here the results from the study of the collection of African Red Slip Ware stored at the Gap Museum. The analyzed ceramics, which came from the Tunisian hinterland, were compared with each other and with known reference materials (Mukaï et al. 2016). Although these analyses were performed on a limited number of specimens, the information exchange between each method led to good results. First of all, the archaeological study allowed determining three main rather homogeneous ARS groups, and one of them was related to a known workshop of Tunisian inland (Sidi Aïch) (Mukaï et al. 2016) (Table 1). Some outlier samples were referred to different unknown productions.

As for the petrographic analysis, two rather homogeneous groups were identified, in good correlation with the archaeological observations, especially concerning Sidi Aïch workshop (Mukaï et al. 2016). However, some samples seem to be in contradiction with the archaeological data.

The geochemical results are only partly in agreement with the archaeological and petrographic ones (Table 1) (Mukaï et al. 2016). Nonetheless, this combined approach is to be verified with an increase of the analyzed samples, including workshops references.

Table 1 Summary of archaeological, petrographical and geochemical data

Inventory	Petrographic interpretation	Portable XRF	Archaeological interpretation
AB 328	Near SA	Near SA	Regional, not SA
AB 376	Near SA	?	Regional, not SA
AB 059	SA not confirmed	Comparable SA	SA
AB 080	SA not confirmed	Near SA	SA
AB 061	SA not confirmed	Comparable SA	Regional, not SA
AB 379	SA not confirmed	Comparable SA	Regional, not SA?
AB 373	SA not confirmed	= SA	Regional, not SA?
AB 062	Regional, not SA	Near SK	Regional, not SA
AB 063	Regional, not SA	Near SK	Regional, not SA
AB 047	Isolated	Near SK	Not regional
AB 028	Near ARS "C/E"	Comparable SA	Not regional
AB 112	Lambaesis	/	Lambaesis

Abbreviations: *SA* Sidi Aïch; *SK* Sidi Khalifa [from (Mukaï et al. 2016)]

4 Discussion

The application of the above discussed multidisciplinary approach on *Thaenae*'s pottery production is particularly interesting, because of the nature of available materials and the refinement possibilities of the measurement protocols.

In the continuity of the test in Gap, the study of the ARS findings at *Thaenae* is an essential point. Indeed, the field surveys revealed an abundance of ceramics imported on the site belonging to ARS categories "C/E" and "E", produced in Djilma (Peacock et al. 1990) in the region of Sbeitla and for which we have reference tables (Mackensen and Schneider 2006). On the basis of this archaeological analysis, the petrographic and geochemical study would make it possible to define whether the tableware present in *Thaenae* comes from local workshops or from other workshops located in or far awy from the same region.

In this way, the characterization of *Thaenae* ceramic production with the construction of mineralogical, petrographic and chemical reference groups is essential. The archaeometric characterization of the workshops, identified in two main areas (Panella 1973; Bonifay 2004), is still preliminary (Capelli et al. 2016) and for some workshops which are better characterized archaeologically and petrographically (Capelli et al. 2016), the XRF characterization will be carried out to compare these productions with other African productions.

As already mentioned, the application of portable XRF on ceramics is still experimental unlike "classical" XRF spectrometers used in many previous studies on African ceramics (Brun 2004, 2007). Portable XRF sometimes gives less precise chemical information in comparison with other XRF analytical methods, but its development will be very useful in archaeology by giving access to museum collections. Like in the case of the Gap collection, this non-destructive method will allow to compare our *Thaenae* findings with complete specimens stored in Sfax museum, discovered in *Thaenae*'s ancient excavations.

5 Conclusion

This first test on Gap's ceramic collection is very promising. Despite some data discrepancies, the analyses confirmed the provenance of the majority of objects. This combined approach will develop protocols and reference systems initiated during previous archaeometric studies and will be used for future multidisciplinary research on other ceramic productions or materials.

In comparison with Northern Tunisia, Late Antiquity economy and ceramic production in South Byzacena are still little known. This research project will also provide important data about economic routes between the internal regions and the coastline and also between Roman Africa and the rest of the Mediterranean.

References

Bonifay, M.: Etude sur la céramique romaine tardive d'Afrique. BAR IS 1301, Oxford (2004)

Bonifay, M., Capelli, C., Brun, C.: Pour une approche intégrée archéologique, pétrographique et géochimique des sigillées africaines. In: Cavalieri, M. (eds.) Industria Apium. L'archéologie: une démarche singulière, des pratiques multiples. Hommages à Raymond Brulet, pp. 41–62. Fervet Opus 1, Louvain (2012)

Brun, C.: Un ensemble de céramiques du IVe s. dans une citerne du capitole. Annexe: Détermination d'origine par fluorescence X de quelques exemplaires de céramiques. In: Ben Hassen, H., Maurin, L. (eds.) Oudhna (Uthina), colonie de vétérans de la XIIIe légion. Histoire, urbanisme, fouilles et mise en valeur des monuments, pp. 245–258. Ausonius-Publications (Mémoires 13), Bordeaux/Paris/Tunis (2004)

Brun, C.: Étude technique des productions de l'atelier de Sidi Khalifa (Pheradi Maius, Tunisie): céramiques culinaires, sigillées et cazettes. In: Bonifay, M., Tréglia, J.-C. (eds.) LRCW 2. Late Roman Coarse Wares, Cooking Wares and Amphorae in the Mediterranean. Archaeology and Archaeometry, pp. 569–579. BAR IS 1662, Oxford (2007)

Capelli, C., Bonifay, M.: Archéométrie et archéologie des céramiques africaines: une approche pluridisciplinaire. In: Bonifay, M., Tréglia, J.-C. (eds.) LRCW 2. Late Roman Coarse Wares, Cooking Wares and Amphorae in the Mediterranean. Archaeology and Archaeometry, pp. 551–567. BAR IS 1662, Oxford (2007)

Capelli, C., Bonifay, M.: Archeologia e archeometria delle anfore dell'Africa romana. Nuovidati e problemi aperti. In: Ferrandes, A., Pardini, G. (eds.) Le regole del gioco. Tracce, archeologi, racconti. Studi in onore di Clementina Panella, pp. 535–557. Lexicon Topographicum Urbis Romae Suppl. VI, Rome (2016)

Frahm, E.: Ceramic studies using portable XRF: from experimental tempered ceramics to imports and imitations at Tell Mozan, Syria. J. Archaeol. Sci. **90**, 13–38 (2018)

Hunt, A.M.W., Speakman, R.J.: Portable XRF analysis of archaeological sediments and ceramics. J. Archaeol. Sci. **53**, 626–638 (2015)

Mackensen, M., Schneider, G.: Production centres of African Red Slip Ware (2nd-3rd c.) in Northern and Central Tunisia. J. Roman Archaeol. **19**, 163–190 (2006)

Mukaï, T., Rêve, R., Bonifay, M., Aibeche, Y., Ambrosi, J.-P., Borgard, P., Capelli, C., Chiaramella, Y., Copetti, A., Durand, C., Foy, D., Nasr, M., Verlinden, F.: Etude de la collection Aubert-Buès d'antiquités africaines au musée de Gap: premiers résultats. Antiquités Africaines **52**, 157–184 (2016)

Panella, C.: Le anfore. In: Carandini, A., Panella, C. (eds.) Ostia III. Le Terme del Nuotatore. Scavo degli ambienti III, VI, VII. Scavo dell'ambiente V e di un saggio nell'area SO, pp. 460–633. Studi Miscellanei 21, Rome (1973)

Peacock, D.P.S., Bejaoui, F., Ben Lazreg, N.: Roman Pottery production in Central Tunisia. J. Roman Archaeol. **3**, 59–84 (1990)

"Giallo Antico" in Roman Architecture of Lombardy: A Preliminary Survey

Roberto Bugini, Luisa Folli, and Romina Marchisio

Abstract

"Giallo antico" is a historical term to identify a limestone coming from Chemtou (Jendouba, northern Tunisia) featuring a yellow ground crossed with red veins. This stone pertains to the ancient coloured marbles, a group of different stone materials used in architecture and for decoration in the whole Roman Empire. Giallo antico was extensively used in Lombardy (Northern Italy) as witnessed in many archaeological sites (Temples, houses, villas, baths and mausoleums) of Milano, Como, Cremona, Brescia. A preliminary survey allowed classifying hundreds of artefacts of Giallo antico (slab for flooring and for veneering, architectural elements as shafts or capitals) occurring also in the rural areas of the regional territory.

Keywords

Giallo antico • Coloured marbles • Roman architecture Tunisia • Lombardy

1 Introduction

The use of coloured marbles in Roman architecture is a significant and original feature of this period and it was spread in the whole territory of the Empire. Coloured marbles, a mix of porphyries, granites, veined marbles, breccias, limestones, alabasters, ophicalcites were employed to make veneers, pavements or decorations exploiting a prodigious variety of patterns. The coloured marbles originated in the whole Mediterranean basin, with a prevalence of the eastern countries (Greece, Turkey, Egypt). The study aimed to emphasize the diffusion of one of these high valued materials, "Giallo antico" from Tunisia, in various parts of Lombardy, a region far away from Rome and from the routes across the sea.

2 Methodology

One of the most expensive ancient coloured marbles was the *Marmor numidicum*, called "Giallo antico" by the stone-cutters of the Papal court in Rome. This marble originates in Tunisia (called *Numidia* by the Romans), in the north-western part near Chemtou village (*Colonia Julia Augusta Numidica Simitthu*), facing the Medjerda river about 20 km west of Jendouba; the western slope of the hill, north of the village, is hewn by huge ancient quarries.

Geologically, the marble is a Triassic limestone with iron hydroxides. The brecciated aspect shows yellow fragments, from dark gold to orange or yellowish white, encircled by veins or bands, from purple red to dark red or brown (Röder 1988; Lazzarini 2004).

Giallo antico was largely employed in Rome; it is witnessed in column shafts (house of Augustus on Palatine hill; Pantheon; Forum of Trajan; Arch of Constantine), slabs for *opus sectile* (Forum of Augustus, Villa Adriana, Pantheon, Ostia), statues (now Copenhagen, Glyptotek), mouldings (Apollo Sosianus), inlays (Horti Lamiani). The marble was spread in northern Italy (from Verona to Aquileja); southern France (from Orange to Marseille), Spain (Córdoba), Tunisia (*Carthago*), Libya (*Leptis Magna*), Turkey (Bergama), and Syria (Palmyra) (Lazzarini 2004).

Different archaeological sites of Lombardy were surveyed in order to identify the Giallo antico artefacts among thousand coloured marble pieces dug out in different excavations. The preliminary survey allows the collection of selected samples (Cremona Marconi; Milano Mausoleum; Brescia Capitolium, Desenzano Villa, Toscolano) that were observed in optical microscopy on thin section. The samples

R. Bugini (✉) · L. Folli
Ist. CNR Conservazione Beni Culturali, Via Cozzi 53, 20125 Milan, Italy
e-mail: bugini@icvbc.cnr.it

R. Marchisio
Dip. Archeologia E Storia Dell'Arte, Università Cattolica, Lgo Gemelli 1, Milan, Italy

H. Chenchouni et al. (eds.), *Exploring the Nexus of Geoecology, Geography, Geoarcheology and Geotourism: Advances and Applications for Sustainable Development in Environmental Sciences and Agroforestry Research*, Advances in Science, Technology & Innovation, https://doi.org/10.1007/978-3-030-01683-8_23

show very fine-grained equi-granular calcite crystals (av. size 0.02 mm), coarser calcite (0.5 mm) aggregates, rectangular crystals of plagioclase (0.4 mm), quartz, undulating veins of iron hydroxides associated with coarse grained calcite crystals (0.5 mm). These features match those of Chemtou quarry samples.

3 Results

3.1 Cremona

- Piazza Marconi (1st century); side by side houses in the southern part of the town. Coloured marbles were dispersed after fires and pillages (year 69 CE). Giallo antico was used for pavements (slab, tile) and veneers (slab, fillet, moulding, small pilaster). An outstanding work is a finely carved leg of a table.
- Via Cadolini (4th century); coloured marbles inserted in a mosaic floor of a *domus* located in the town centre. Broken slabs (5–15 cm long and 1–2 cm thick) of Giallo antico, Pavonazzetto, Portasanta, Alabasters with Africano, Rosso antico, Semesanto. Each slab was encircled by two rows of tesserae of white limestone.

3.2 Milano—Como

- Via Brisa (3rd century): a building pertaining to the Roman imperial palace contains slabs and polygonal tiles of Giallo antico (0.65–1.97 cm thick) together with Fior di Pesco, Portasanta, Pavonazzetto, Cipollino mandolato and red and green porphyries; moulded slabs of Cipollino; capitals of Rosso antico.
- Via Correnti (mid 1st Century): a mortar pavement (*domus* off of the urban walls in the south-western part of the town) contains concentric lines of polygonal slabs of coloured marbles (Giallo antico, Africano, Settebasi, Cipollino, Pavonazzetto, Porfido rosso, Portasanta) together with cubic tesserae of black limestone.
- Via Gorani and via San Protaso (4th century): mosaic floors of two buildings, pertaining to the Roman imperial palace, include coloured marble as broken slabs; each slab (4–6 cm long) is surrounded by some rows of white limestone tesserae. Giallo antico is laid with Africano, Alabaster, Breccia corallina and Portasanta.
- Via Moneta (2nd century): an indistinct layer of debris located very close to the Mint and the Forum. Slabs of Giallo antico (minimum thickness 0.48 cm) and Cipollino, Fior di Pesco, Pavonazzetto, red and green porphyries, Africano, Alabaster, Breccia corallina, Portasanta, Rosso antico, Settebasi, Verde antico.
- Herculean baths (late 3rd century): a huge building (sides about 100 m long) on squared plan, then completely disappeared under the modern town. Giallo antico was used in rectangular slabs for wall veneers or floor tiles. The thickness varies from 0.5 mm to 1.5 cm for the thinner slabs and from 1 to 5 cm from the thicker ones. Twenty two sculpted elements, pertaining to the shaft of spiral fluting columns with sharp or blunt edges, are noticeable; the original size of each shaft ranged between 2.8 and 3.0 m with a diameter of 0.35 m.
- Imperial Mausoleum (4th century): the octagonal edifice, built off of the urban walls in the south-west of the town, contains fillets and slabs of Giallo antico used or reused for wall veneers together with Cipollino, red or green porphyries.
- Como, via Parini (2nd century): a layer of debris under a road surface, close to an urban tower. Slabs, fillets and mouldings of Giallo antico (slab 0.3 cm thick) and Africano, Alabaster, Cipollino, Fior di Pesco, Pavonazzetto, Rosso antico.

3.3 Brescia

- Capitolium: an upraised colonnaded complex located on the northern side of the Forum and erected by Vespasianus after the victorious war of Three Emperors (year 69 CE). The floor of the main *aula* was made using rectangular slabs of Giallo antico (60 cm long, 30 cm wide and 0.15 cm thick) and Pavonazzetto.
- Martinengo-Cesaresco, via Musei 32 (2nd century): a bath now confined in the 18th century palace contains slabs and mouldings of Giallo antico with Africano, Alabaster, Breccia corallina, Cipollino, Pavonazzetto, Settebasi, Verde antico.
- Desenzano del Garda (4th century): a residential villa located on the southern shores of lake Garda contains mouldings and slabs of Giallo antico (thickness 0.75–3.25 cm) with Africano, Greco scritto, Pavonazzetto, Verde antico and others.
- Desenzano, Faustinella (first half 4th century): a rustic complex located in the plain: slabs (0.47–1.9 cm thick) and mouldings of Giallo antico with Africano, Cipollino, Porfido serpentino verde, Portasanta, Rosso antico, Settebasi.
- Toscolano (Brescia) (2nd century): a residential villa on the western banks of the mid lake Garda contains slabs and mouldings of Giallo antico (thinner slab 0.70 cm thick) and Africano, Cipollino, Palombino, Porfido rosso Antico, Porfido serpentino verde, Portasanta, Settebasi.

4 Discussion

Giallo antico was one of the most used coloured marbles in Lombardy, a region split, in the Roman period, into two Regions: *Venetia* and *Transpadana*. It was common in public or religious buildings, houses, residential or rustic villas. This use seems to start in the mid of the 1st century CE in Milano and the reuse, a practice noted since the 4th century, makes vague the end (Bugini and Folli 2005). Marbles were shipped to Ostia and along Italian coasts to the mouth of river Po (Adriatic Sea), then reaching the Lombard territory through waterways (Maischberger 1997).

Starting from the Middle Ages, Giallo antico was replaced by two Italian lithotypes: Giallo di Siena (marble, Jurassic, Montagnola Senese) and Giallo Torri (nodular limestone, Jurassic, lake Garda). Other yellow Mesozoic limestones were used in Baroque churches of Naples (Mondragone) and Palermo (Castronovo).

The quotations of ancient authors do not allow us to understand the chronology of the use of this marble and they reveal uncertain information on Giallo antico and its provenance. First hand notices seem to be written by Pliny (*Naturalis Historia*, 36–49) about the use during M. Lepidus consulate (year 78 BCE) as door-sill and not in the form of columns or slabs. Modern Italian or French authors mainly ignored the Giallo antico. Agostino del Riccio is the sole eyewitness of the reuse in some churches of Florence. Italian marble treatises (19th century) listed different varieties of yellow marbles, without any information on their provenance (Gnoli 1989).

5 Conclusions

Giallo antico elements were found in Roman buildings throughout the territory of Lombardy from Milano to Como, Brescia and lake Garda. This fine-grained yellow limestone, coming from Tunisia, was used to make different artefacts. So, the textural homogeneity was exploited to obtain very thin slabs (0.3 cm) used for inlays and fine carving used for spiral columns, capitals or legs of tables.

Giallo antico artefacts were used not only in Milan or other important towns, but also present in rural areas, far from the main routes. The chronology starts from mid 1st century and attains the acme in the 3rd century; then coloured marbles were reused in some Medieval buildings. Giallo antico was also replaced, since the Middle Ages, by yellow marbles from Sienna or yellow limestones from lake Garda (Verona). Giallo antico was always joined with the recurrent group of coloured marbles of the Mediterranean basin and with white ones.

References

Bugini, R., Folli, L.: Sull'uso di marmi colorati (Lombardia). Marmora **1**, 145–168 (2005)
Gnoli, R.: Marmora romana, 2nd edn. Dell'Elefante, Roma (1989)
Lazzarini, L.: Pietre e marmi antichi. Cedam, Padova (2004)
Maischberger, M.: Marmor in Rome. Reichert, Wiesbaden (1997)
Röder, G.: Numidian marble. In Herz, N. (ed.) Classical Marbles, pp. 91–96 (1988)

Printed by Printforce, the Netherlands